MW01079626

Brain Facts

A PRIMER ON THE BRAIN AND NERVOUS SYSTEM

THE SOCIETY FOR NEUROSCIENCE

THE SOCIETY FOR NEUROSCIENCE

The Society for Neuroscience is the world's largest organization of scientists and physicians dedicated to understanding the brain, spinal cord and peripheral nervous system.

Neuroscientists investigate the molecular and cellular levels of the nervous system; the neuronal systems responsible for sensory and motor function; and the basis of higher order processes, such as cognition and emotion. This research provides the basis for understanding the medical fields that are concerned with treating nervous system disorders. These medical specialties include neurology, neurosurgery, psychiatry and ophthalmology.

Founded in 1970, the Society has grown from 500 charter members to more than 29,000 members. Regular members are residents of Canada, Mexico and the United States—where more than 100 chapters organize local activities. The Society's membership also includes many scientists from throughout the world, particularly Europe and Asia.

The purposes of the Society are to:

- Advance the understanding of the nervous system by bringing together scientists from various backgrounds and by encouraging research in all aspects of neuroscience.
- Promote education in the neurosciences.
- Inform the public about the results and implications of new research.

The exchange of scientific information occurs at an annual fall meeting that presents more than 14,000 reports of new scientific findings and includes more than 25,000 participants. This meeting, the largest of its kind in the world, is *the* arena for the presentation of new results in neuroscience.

The Society's bimonthly journal, *The Journal of Neuroscience,* contains articles spanning the entire range of neuroscience research and has subscribers worldwide. A series of courses, workshops and symposia held at the annual meeting promote the education of Society members. The *Neuroscience Newsletter* informs members about Society activities.

A major mission of the Society is to inform the public about the progress and benefits of neuroscience research. The Society provides information about neuroscience to school teachers and encourages its members to speak to young people about the human brain and nervous system.

Brain Facts

Introduction

It sets humans apart from all other species by allowing us to achieve the wonders of walking on the moon and composing masterpieces of literature, art and music. Throughout recorded time, the human brain—a spongy, three-pound mass of fatty tissue—has been compared to a telephone switchboard and a supercomputer.

But the brain is much more complicated than any of these devices, a fact scientists confirm almost daily with each new discovery. The extent of the brain's capabilities is unknown, but it is the most complex living structure known in the universe.

This single organ controls all body activities, ranging from heart rate and sexual function to emotion, learning and memory. The brain is even thought to influence the response to disease of the immune system and to determine, in part, how well people respond to medical treatments. Ultimately, it shapes our thoughts, hopes, dreams and imagination. In short, the brain is what makes us human.

Neuroscientists have the daunting task of deciphering the mystery of this most complex of all machines: how as many as a trillion nerve cells are produced, grow and organize themselves into effective, functionally active systems that ordinarily remain in working order throughout a person's lifetime.

The motivation of researchers is twofold: to understand human behavior better—from how we learn to why people have trouble getting along together—and to discover ways to prevent or cure many devastating brain disorders.

The more than 1,000 disorders of the brain and nervous system result in more hospitalizations than any other disease group, including heart disease and cancer. Neurological illnesses affect more than 50 million Americans annually at costs exceeding $400 billion. In addition, mental disorders, excluding drug and alcohol problems, strike 44 million adults a year at a cost of some $148 billion.

However, during the congressionally designated Decade of the Brain, which ended in 2000, neuroscience made significant discoveries in these areas:

- Genetics. Key disease genes were identified that underlie several neurodegenerative disorders—including Alzheimer's disease, Huntington's disease, Parkinson's disease and amyotrophic lateral sclerosis. This has provided new insights into underlying disease mechanisms and is beginning to suggest new treatments.

With the mapping of the human genome, neuroscientists will be able to make more rapid progress in identifying genes that either contribute to human neurological disease or that directly cause disease. Mapping animal genomes will aid the search for genes that regulate and control many complex behaviors.

- Brain Plasticity. Scientists began to uncover the molecular bases of neural plasticity, revealing how learning and memory occur and how declines might be reversed. It also is leading to new approaches to the treatment of chronic pain.
- New Drugs. Researchers gained new insights into the mechanisms of molecular neuropharmacology, which provides a new understanding of the mechanisms of addiction. These advances also have led to new treatments for depression and obsessive-compulsive disorder.
- Imaging. Revolutionary imaging techniques, including magnetic resonance imaging and positron emission tomography, now reveal brain systems underlying attention, memory and emotions and indicate dynamic changes that occur in schizophrenia.
- Cell Death. The discovery of how and why neurons die, as well as the discovery of stem cells, which divide and form new neurons, has many clinical applications. This has dramatically improved the outlook for reversing the effects of injury both in the brain and spinal cord. The first effective treatments for stroke and spinal cord injury based on these advances have been brought to clinical practice.
- Brain Development. New principles and molecules responsible for guiding nervous system development now give scientists a better understanding of certain disorders of childhood. Together with the discovery of stem cells, these advances are pointing to novel strategies for helping the brain or spinal cord regain functions lost to diseases.

Federal neuroscience research funding of more than $4 billion annually and private support should vastly expand our knowledge of the brain in the years ahead.

This book only provides a glimpse of what is known about the nervous system, the disorders of the brain and some of the exciting avenues of research that promise new therapies for many neurological diseases.

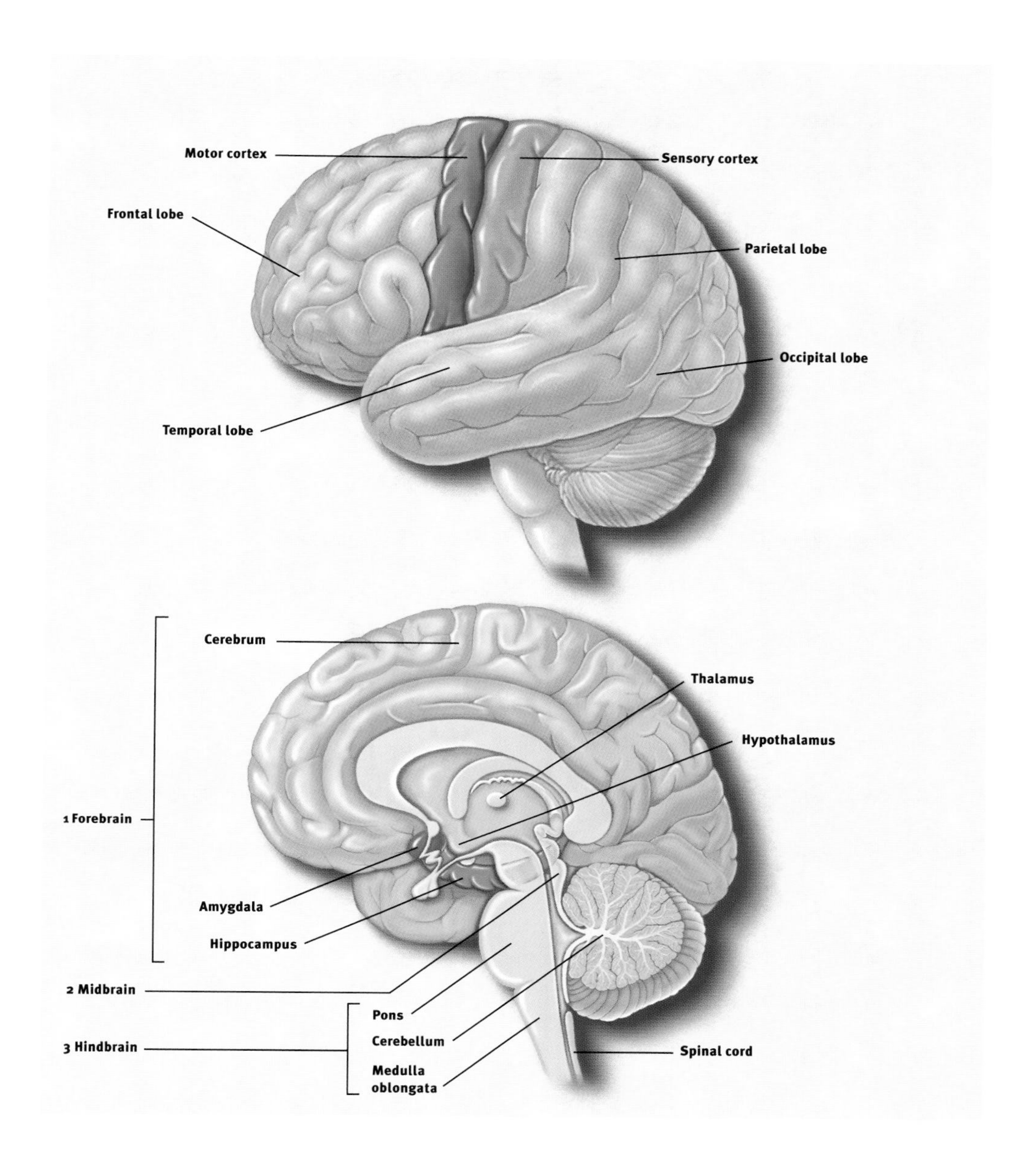

THE BRAIN. Cerebral cortex (above). This part of the brain is divided into four sections: the occipital lobe, the temporal lobe, the parietal lobe and the frontal lobe. Functions, such as vision, hearing and speech, are distributed in selected regions. Some regions are associated with more than one function. Major internal structures (below). The (1) forebrain is credited with the highest intellectual functions—thinking, planning and problem-solving. The hippocampus is involved in memory. The thalamus serves as a relay station for almost all of the information coming into the brain. Neurons in the hypothalamus serve as relay stations for internal regulatory systems by monitoring information coming in from the autonomic nervous system and commanding the body through those nerves and the pituitary gland. On the upper surface of the (2) midbrain are two pairs of small hills, colliculi, collections of cells that relay specific sensory information from sense organs to the brain. The (3) hindbrain consists of the pons and medulla oblongata, which help control respiration and heart rhythms, and the cerebellum, which helps control movement as well as cognitive processes that require precise timing.

THE TOLL OF SELECTED BRAIN AND NERVOUS SYSTEM DISORDERS*

Condition	*Total Cases*	*Costs Per Year*
Hearing Loss	28 million	$ 56 billion
All Depressive Disorders	18.8 million	$ 44 billion
Alzheimer's Disease	4 million	$ 100 billion
Stroke	4 million	$ 30 billion
Schizophrenia	3 million	$ 32.5 billion
Parkinson's Disease	1.5 million	$ 15 billion
Traumatic Head Injury	1 million	$ 48.3 billion
Multiple Sclerosis	350,000	$ 7 billion
Spinal Cord Injury	250,000	$ 10 billion

* *Estimates provided by the National Institutes of Health and voluntary organizations.*

The Neuron

A specialized cell designed to transmit information to other nerve cells, muscle or gland cells, the neuron is the basic working unit of the brain. The brain is what it is because of the structural and functional properties of neurons. The brain contains between one billion and one trillion neurons.

The neuron consists of a *cell body* containing the nucleus and an electricity-conducting fiber, the *axon*, which also gives rise to many smaller axon branches before ending at *nerve terminals*. *Synapses*, from the Greek words meaning to "clasp together," are the contact points where one neuron communicates with another. Other cell processes, *dendrites*, Greek for the branches of a tree, extend from the neuron cell body and receive messages from other neurons. The dendrites and cell body are covered with synapses formed by the ends of axons of other neurons.

Neurons signal by transmitting electrical impulses along their axons that can range in length from a tiny fraction of an inch to three or more feet. Many axons are covered with a layered insulating *myelin* sheath, made of specialized cells, that speeds the transmission of electrical signals along the axon.

Nerve impulses involve the opening and closing of *ion channels*, water-filled molecular tunnels that pass through the cell membrane and allow ions—electrically charged atoms—or small molecules to enter or leave the cell. The flow of these ions creates an electrical current that produces tiny voltage changes across the membrane.

The ability of a neuron to fire depends on a small difference in electrical charge between the inside and outside of the cell. When a nerve impulse begins, a dramatic reversal occurs at one point on the cell's membrane. The change, called an *action potential*, then passes along the membrane of the axon at speeds up to several hundred miles an hour. In this way, a neuron may be able to fire impulses scores or even hundreds of times every second.

On reaching the ends of an axon, these voltage changes trigger the release of *neurotransmitters*, chemical messengers. Neurotransmitters are released at nerve ending terminals and bind to receptors on the surface of the target neuron.

These receptors act as on and off switches for the next cell. Each receptor has a distinctly shaped part that exactly matches a particular chemical messenger. A neurotransmitter fits into this region in much the same way as a key fits into an automobile ignition. And when it does, it alters the neuron's outer membrane and triggers a change, such as the contraction of a muscle or increased activity of an enzyme in the cell.

Knowledge of neurotransmitters in the brain and the action of drugs on these chemicals—gained largely through the study of animals—is one of the largest fields in neuroscience. Armed with this information, scientists hope to understand the circuits responsible for disorders such as Alzheimer's disease and Parkinson's disease. Sorting out the various chemical circuits is vital to understanding how the brain stores memories, why sex is such a powerful motivation and what is the biological basis of mental illness.

Neurotransmitters

Acetylcholine The first neurotransmitter to be identified 70 years ago, was acetylcholine (ACh). This chemical is released by neurons connected to voluntary muscles (causing them to contract) and by neurons that control the heartbeat. ACh also serves as a transmitter in many regions of the brain.

ACh is formed at the axon terminals. When an action potential arrives at the terminal, the electrically charged calcium ion rushes in, and ACh is released into the synapse and attaches to ACh receptors. In voluntary muscles, this opens sodium channels and causes the muscle to contract. ACh is then broken down and re-synthesized in the nerve terminal. Antibodies that block the receptor for ACh cause *myasthenia gravis*, a disease characterized by fatigue and muscle weakness.

Much less is known about ACh in the brain. Recent discoveries suggest, however, that it may be critical for normal attention, memory and sleep. Since ACh-releasing neurons die in Alzheimer's patients, finding ways to restore this neurotransmitter is one goal of current research.

Amino Acids Certain amino acids, widely distributed throughout the body and the brain, serve as the building blocks of proteins. However, it is now apparent that certain amino acids can also serve as neurotransmitters in the brain.

The neurotransmitters *glutamate* and *aspartate* act as excitatory signals. *Glycine* and *gamma-aminobutyric acid* (GABA) inhibit the firing of neurons. The activity of GABA is increased by *benzodiazepine* (Valium) and by anticonvulsant drugs. In Huntington's disease, a hereditary disorder that begins during mid-life, the GABA-producing neurons in the brain centers coordinating movement degenerate, thereby causing incontrollable movements.

Glutamate or aspartate activate *N-methyl-D-aspartate* (NMDA) receptors, which have been implicated in activities ranging from learning and memory to development and specification of nerve contacts in a developing animal. The stimulation of NMDA receptors may promote beneficial changes in the brain, whereas overstimulation can cause nerve cell damage or cell death in trauma and stroke.

Key questions remain about this receptor's precise structure, regulation, location and function. For example, developing drugs to block or stimulate activity at NMDA receptors holds

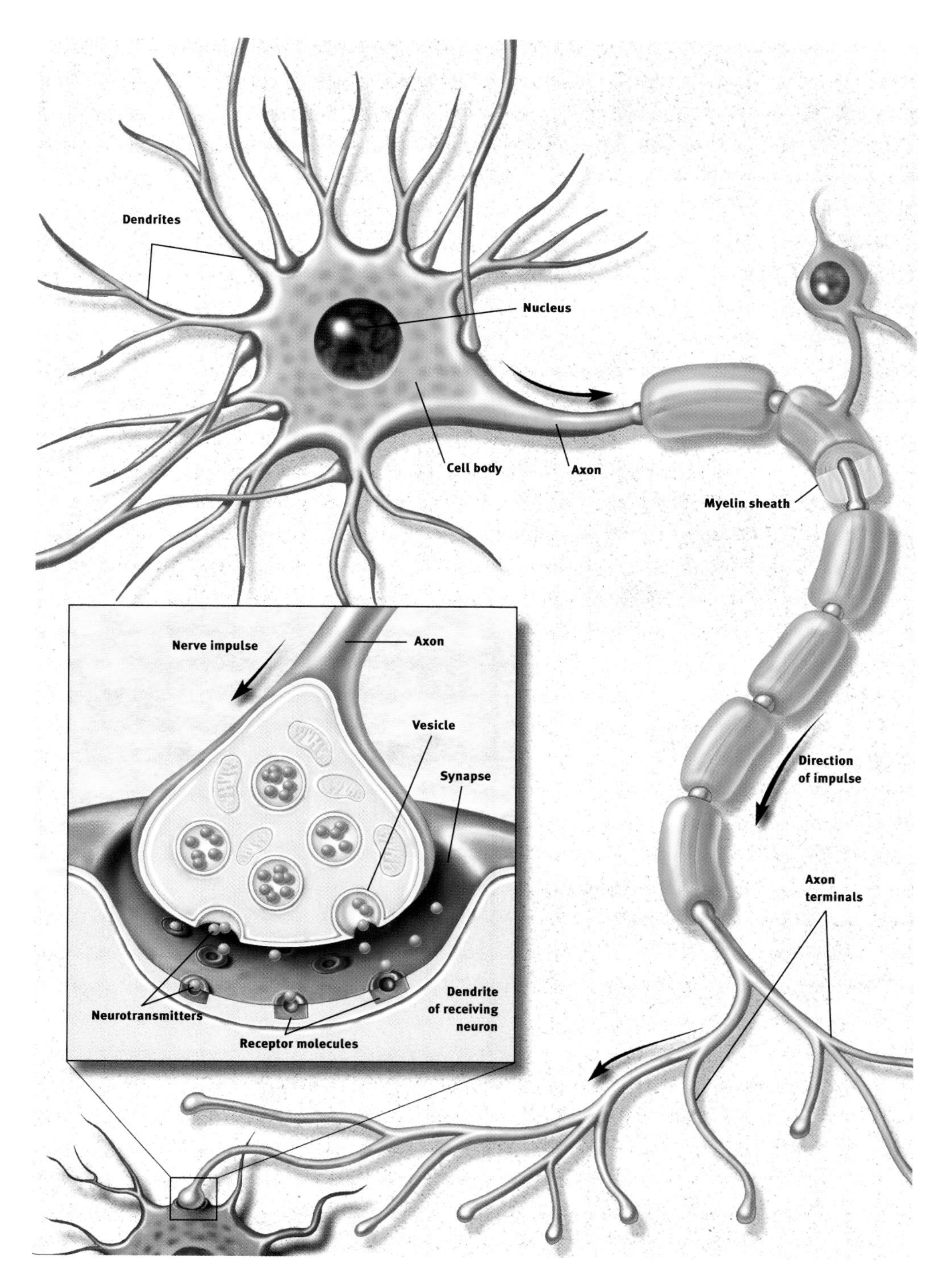

NEURON. A neuron fires by transmitting electrical signals along its axon. When signals reach the end of the axon, they trigger the release of neurotransmitters that are stored in pouches called vesicles. Neurotransmitters bind to receptor molecules that are present on the surfaces of adjacent neurons. The point of virtual contact is known as the synapse.

promise for improving brain function and treating neurological disorders. But this work is still in the early stage.

Catecholamines *Dopamine* and *norepinephrine* are widely present in the brain and peripheral nervous system. Dopamine, which is present in three circuits in the brain, controls movement, causes psychiatric symptoms such as psychosis and regulates hormonal responses.

The dopamine circuit that regulates movement has been directly related to disease. The brains of people with *Parkinson's disease*—with symptoms of muscle tremors, rigidity and difficulty in moving—have practically no dopamine. Thus, medical scientists found that the administration of *levodopa*, a substance from which dopamine is synthesized, is an effective treatment for Parkinson's, allowing patients to walk and perform skilled movements successfully.

Another dopamine circuit is thought to be important for cognition and emotion; abnormalities in this system have been implicated in schizophrenia. Because drugs that block dopamine receptors in the brain are helpful in diminishing psychotic symptoms, learning more about dopamine is important to understanding mental illness.

In a third circuit, dopamine regulates the endocrine system. It directs the hypothalamus to manufacture hormones and hold them in the pituitary gland for release into the bloodstream, or to trigger the release of hormones held within cells in the pituitary.

Nerve fibers containing norepinephrine are present throughout the brain. Deficiencies in this transmitter occur in patients with Alzheimer's disease, Parkinson's disease and those with *Korsakoff's syndrome*, a cognitive disorder associated with chronic alcoholism. Thus, researchers believe norepinephrine may play a role in both learning and memory. Norepinephrine also is secreted by the sympathetic nervous system in the periphery to regulate heart rate and blood pressure. Acute stress increases the release of norepinephrine.

Serotonin This neurotransmitter is present in many tissues, particularly blood platelets and the lining of the digestive tract and the brain. Serotonin was first thought to be involved in high blood pressure because it is present in blood and induces a very powerful contraction of smooth muscles. In the brain, it has been implicated in sleep, mood, depression and anxiety. Because serotonin controls the different switches affecting various emotional states, scientists believe these switches can be manipulated by analogs, chemicals with molecular structures similar to serotonin. Drugs that alter serotonin's action, such as *fluoxetine* (Prozac), have relieved symptoms of depression and obsessive-compulsive disorder.

Peptides These chains of amino acids linked together, have been studied as neurotransmitters only in recent years. Brain peptides called *opioids* act like opium to kill pain or cause sleepiness. (Peptides differ from proteins, which are much larger and more complex combinations of amino acids.)

In 1973, scientists discovered receptors for opiates on neurons in several regions in the brain that suggested the brain must make substances very similar to opium. Shortly thereafter, scientists made their first discovery of an opiate produced by the brain that resembles morphine, an opium derivative used medically to kill pain. They named it *enkephalin*, literally meaning "in the head." Subsequently, other opiates known as *endorphins*—from endogenous morphine—were discovered.

The precise role of the opioids in the body is unclear. A plausible guess is that enkephalins are released by brain neurons in times of stress to minimize pain and enhance adaptive behavior. The presence of enkephalins may explain, for example, why injuries received during the stress of combat often are not noticed until hours later.

Opioids and their receptors are closely associated with pathways in the brain that are activated by painful or tissue-damaging stimuli. These signals are transmitted to the *central nervous system*—the brain and spinal cord—by special sensory nerves, small myelinated fibers and tiny unmyelinated or *C fibers*.

Scientists have discovered that some C fibers contain a peptide called *substance P* that causes the sensation of burning pain. The active component of chili peppers, capsaicin, causes the release of substance P.

Trophic factors Researchers have discovered several small proteins in the brain that are necessary for the development, function and survival of specific groups of neurons. These small proteins are made in brain cells, released locally in the brain, and bind to receptors expressed by specific neurons. Researchers also have identified genes that code for receptors and are involved in the signaling mechanisms of trophic factors. These findings are expected to result in a greater understanding of how trophic factors work in the brain. This information also should prove useful for the design of new therapies for brain disorders of development and for degenerative diseases, including Alzheimer's disease and Parkinson's disease.

Hormones After the nervous system, the *endocrine system* is the second great communication system of the body. The pancreas, kidney, heart and adrenal gland are sources of hormones. The endocrine system works in large part through the pituitary that secretes hormones into the blood. Because endorphins are released from the pituitary gland into the bloodstream, they might also function as endocrine hormones. Hormones activate specific receptors in target organs that release other hormones into the blood, which then act on other tissues, the pituitary itself and the brain. This system is very important for the activation and control of basic behavioral activities such as sex, emotion, response to stress and the regulation of body functions, such as growth, energy use and metabolism. Actions of hormones show the brain to be very malleable and capable of responding to environmental signals.

The brain contains receptors for both the thyroid hormone and the six classes of steroid hormones—*estrogens, androgens, progestins, glucocorticoids, mineralocorticoids* and *vitamin D*. The receptors are found in selected populations of neurons in the brain and relevant organs in the body. Thyroid and steroid hormones bind to receptor proteins that in turn bind to the DNA genetic material and regulate action of genes. This can result in long-lasting changes in cellular structure and function.

In response to stress and changes in our *biological clocks*, such as day-and-night cycles and jet-lag, hormones enter the blood and travel to the brain and other organs. In the brain, they alter the production of gene products that participate in synaptic neurotransmission as well as the structure of brain cells. As a result, the circuitry of the brain and its capacity for neurotransmission are changed over a course of hours to days. In this way, the brain adjusts its performance and control of behavior in response to a changing environment. Hormones are important agents of protection and adaptation, but stress and stress hormones also can alter brain function, including learning. Severe and prolonged stress can cause permanent brain damage.

Reproduction is a good example of a regular, cyclic process driven by circulating hormones: The hypothalamus produces *gonadotropin-releasing hormone* (GnRH), a peptide that acts on cells in the pituitary. In both males and females, this causes two hormones—the *follicle-stimulating hormone* (FSH) and the *luteinizing hormone* (LH)—to be released into the bloodstream. In males, these hormones are carried to receptors on cells in the testes where they release the male hormone testosterone into the bloodstream. In females, FSH and LH act on the ovaries and cause the release of the female hormones estrogen and progesterone. In turn, the increased levels of testosterone in males and estrogen in females act back on the hypothalamus and pituitary to decrease the release of FSH and LH. The increased levels also induce changes in cell structure and chemistry that lead to an increased capacity to engage in sexual behavior.

Scientists have found statistically and biologically significant differences between the brains of men and women that are similar to sex differences found in experimental animals. These include differences in the size and shape of brain structures in the hypothalamus and the arrangement of neurons in the cortex and hippocampus. Some functions can be attributed to these sex differences, but much more must be learned in terms of perception, memory and cognitive ability. Although differences exist, the brains of men and women are more similar than they are different.

Recently, several teams of researchers have found anatomical differences between the brains of heterosexual and homosexual men. Research suggests that hormones and genes act early in life to shape the brain in terms of sex-related differences in structure and function, but scientists still do not have a firm grip on all the pieces of this puzzle.

Gases Very recently, scientists identified a new class of neurotransmitters that are gases. These molecules—*nitric oxide* and *carbon monoxide*—do not obey the "laws" governing neurotransmitter behavior. Being gases, they cannot be stored in any structure, certainly not in synaptic storage structures. Instead, they are made by enzymes as they are needed. They are released from neurons by diffusion. And rather than acting at receptor sites, they simply diffuse into adjacent neurons and act upon chemical targets, which may be enzymes.

Though only recently characterized, nitric oxide has already been shown to play important roles. For example, nitric oxide neurotransmission governs erection in neurons of the penis. In nerves of the intestine, it governs the relaxation that contributes to normal movements of digestion. In the brain, nitric oxide is the major regulator of the intracellular messenger molecule—*cyclic GMP*. In conditions of excess glutamate release, as occurs in stroke, neuronal damage following the stroke may be attributable in part to nitric oxide. Exact functions for carbon monoxide have not yet been shown.

Second messengers

Recently recognized substances that trigger biochemical communication within cells, second messengers may be responsible for long-term changes in the nervous system. They convey the chemical message of a neurotransmitter (the first messenger) from the cell membrane to the cell's internal biochemical machinery. Second messengers take anywhere from a few milliseconds to minutes to transmit a message.

An example of the initial step in the activation of a second messenger system involves *adenosine triphosphate* (ATP), the chemical source of energy in cells. ATP is present throughout the cell. For example, when norepinephrine binds to its receptors on the surface of the neuron, the activated receptor binds G-proteins on the inside of the membrane. The activated G-protein causes the enzyme *adenylyl cyclase* to convert ATP to *cyclic adenosine monophosphate* (cAMP). The second messenger, cAMP, exerts a variety of influences on the cell, ranging from changes in the function of ion channels in the membrane to changes in the expression of genes in the nucleus, rather than acting as a messenger between one neuron and another. cAMP is called a second messenger because it acts after the first messenger, the transmitter chemical, has crossed the synaptic space and attached itself to a receptor.

Second messengers also are thought to play a role in the manufacture and release of neurotransmitters, intracellular movements, carbohydrate metabolism in the *cerebrum*—the largest part of the brain consisting of two hemispheres—and the processes of growth and development. Direct effects of these substances on the genetic material of cells may lead to long-term alterations of behavior.

Brain development

Three to four weeks after conception, one of the two cell layers of the gelatin-like human embryo, now about one-tenth of an inch long, starts to thicken and build up along the middle. As this flat neural plate grows, parallel ridges, similar to the creases in a paper airplane, rise across its surface. Within a few days, the ridges fold in toward each other and fuse to form the hollow neural tube. The top of the tube thickens into three bulges that form the hindbrain, midbrain and forebrain. The first signs of the eyes and then the hemispheres of the brain appear later.

How does all this happen? Although many of the mechanisms of human brain development remain secrets, neuroscientists are beginning to uncover some of these complex steps through studies of the roundworm, fruit fly, frog, zebrafish, mouse, rat, chicken, cat and monkey.

Many initial steps in brain development are similar across species, while later steps are different. By studying these similarities and differences, scientists can learn how the human brain develops and how brain abnormalities, such as mental retardation and other brain disorders, can be prevented or treated.

Neurons are initially produced along the central canal in the neural tube. These neurons then migrate from their birthplace to a final destination in the brain. They collect together to form each of the various brain structures and acquire specific ways of transmitting nerve messages. Their processes, or axons, grow long distances to find and connect with appropriate partners, forming elaborate and specific circuits. Finally, sculpting action eliminates redundant or improper connections, honing the specificity of the circuits that remain. The result is the creation of a precisely elaborated adult network of 100 billion neurons capable of a body movement, a perception, an emotion or a thought.

Knowing how the brain is put together is essential for understanding its ability to reorganize in response to external influences or to injury. These studies also shed light on brain functions, such as learning and memory. Brain diseases, such as schizophrenia and mental retardation, are thought to result from a failure to construct proper connections during development. Neuroscientists are beginning to discover some general principles to understand the processes of development, many of which overlap in time.

Birth of neurons and brain wiring

The embryo has three primary layers that undergo many interactions in order to evolve into organ, bone, muscle, skin or

BRAIN DEVELOPMENT. The human brain and nervous system begin to develop at three weeks' gestation as the closing neural tube (left). By four weeks, major regions of the human brain can be recognized in primitive form, including the forebrain, midbrain, hindbrain, and optic vesicle (from which the eye develops). Irregular ridges, or convolutions, are clearly seen by six months.

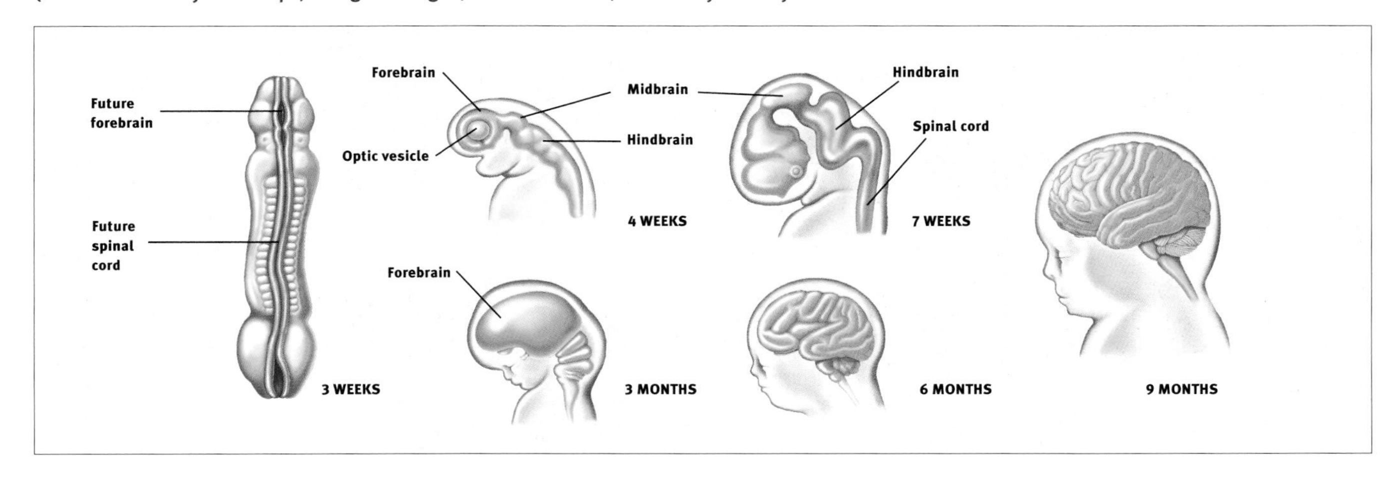

neural tissue. The skin and neural tissue arise from a single layer, known as the ectoderm, in response to signals provided by an adjacent layer, known as the mesoderm.

A number of molecules interact to determine whether the ectoderm becomes neural tissue or develops in another way to become skin. Studies of spinal cord development in frogs show that one major mechanism depends on specific molecules that inhibit the activity of various proteins. If nothing interrupts the activity of such proteins, the tissue becomes skin. If other molecules, which are secreted from mesodermal tissue, block protein signaling, then the tissue becomes neural.

Once the ectodermal tissue has acquired its neural fate, another series of signaling interactions determine the type of neural cell to which it gives rise. The mature nervous system contains a vast array of cell types, which can be divided into two main categories: the neurons, primarily responsible for signaling, and supporting cells called glial cells.

Researchers are finding that the destiny of neural tissue depends on a number of factors, including position, that define the environmental signals to which the cells are exposed. For example, a key factor in spinal cord development is a secreted protein called sonic hedgehog that is similar to a signaling protein found in flies. The protein, initially secreted from mesodermal tissue lying beneath the developing spinal cord, marks young neural cells that are directly adjacent to become a specialized class of glial cells. Cells further away are exposed to lower concentrations of sonic hedgehog protein, and they become the motor neurons that control muscles. An even lower concentration promotes the formation of interneurons that relay messages to other neurons, not muscles.

A combination of signals also determines the type of chemical messages, or neurotransmitters, that a neuron will use to communicate with other cells. For some, such as motor neurons, the choice is invariant, but for others it is a matter of choice. Scientists found that when certain neurons are maintained in a dish without any other cell type, they produce the neurotransmitter norepinephrine. In contrast, if the same neurons are maintained with other cells, such as cardiac or heart tissue cells, they produce the neurotransmitter acetylcholine. Since all neurons have genes containing the information for the production of these molecules, it is the turning on of a particular set of genes that begins the production of specific neurotransmitters. Many researchers believe that the signal to engage the gene and, therefore, the final determination of the chemical messengers that a neuron produces, is influenced by factors coming from the targets themselves.

As neurons are produced, they move from the neural tube's ventricular zone, or inner surface, to near the border of the marginal zone, or the outer surface. After neurons stop dividing, they form an intermediate zone where they gradually accumulate as the brain develops.

The migration of neurons occurs in most structures of the brain, but is particularly prominent in the formation of a large cerebral cortex in primates, including humans. In this structure,

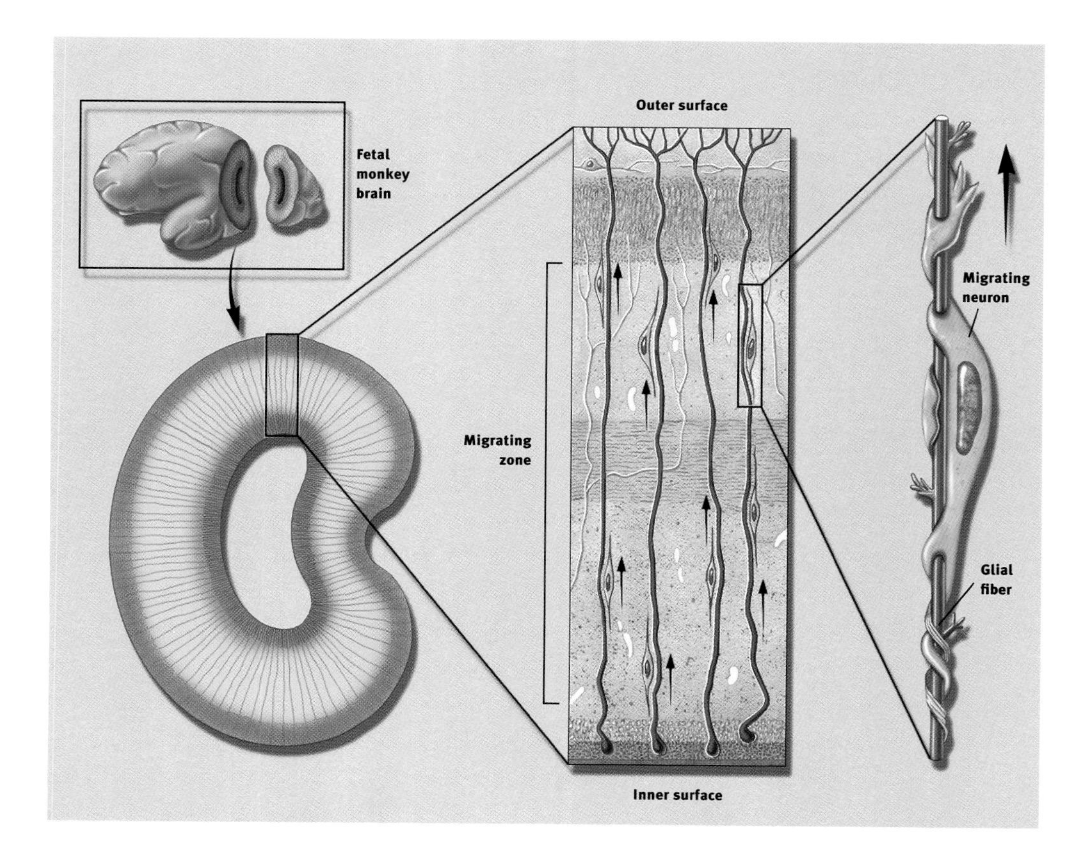

NEURON MIGRATION. A cross-sectional view of the occipital lobe (which processes vision) of a three-month-old monkey fetus brain (center) shows immature neurons migrating along glial fibers. These neurons make transient connections with other neurons before reaching their destination. A single migrating neuron, shown about 2,500 times its actual size (right), uses a glial fiber as a guiding scaffold. To move, it needs adhesion molecules, which recognize the pathway, and contractile proteins to propel it along.

neurons slither from the place of origin near the ventricular surface along nonneuronal fibers that form a trail to their proper destination. Proper neuron migration requires multiple mechanisms, including the recognition of the proper path and the ability to move long distances. One such mechanism for long distance migration is the movement of neurons along elongated fibers that form transient scaffolding in the fetal brain. Many external forces, such as alcohol, cocaine or radiation, prevent proper neuronal migration and result in misplacement of cells, which may lead to mental retardation and epilepsy. Furthermore, mutations in genes that regulate migration have recently been shown to cause some rare genetic forms of retardation and epilepsy in humans.

Once the neurons reach their final location, they must make the proper connections for a particular function, such as vision or hearing, to occur. They do this through their axons. These stalk-like appendages can stretch out a thousand times longer than the cell body from which they arise. The journey of most axons ends when they meet the branching areas, called dendrites, on other neurons. These target neurons can be located at a considerable distance, sometimes at opposite sides of the brain. In the case of a motor neuron, the axon may travel from the spinal cord all the way down to a foot muscle. The linkup sites, called synapses, are where messages are transferred from one neuron in a circuit to the next.

Axon growth is spearheaded by growth cones. These enlargements of the axon's tip actively explore the environment as they seek out their precise destinations. Researchers have discovered that many special molecules help guide growth cones. Some molecules lie on the cells that growth cones contact, while others are released from sources found near the growth cone. The growth cones, in turn, bear molecules that serve as receptors for the environmental cues. The binding of particular signals with its receptors tells the growth cone whether to move forward, stop, recoil or change direction.

Recently researchers have identified some of the molecules that serve as cues and receptors. These molecules include proteins with names such as cadherin, netrin, semaphorin, ephrin, neuropilin and plexin. In most cases, these are families of related molecules; for example there are at least 15 semapohorins and at least 10 ephrins. Perhaps the most remarkable result is that most of these are common to worms, insects and mammals, including humans. Each family is smaller in flies or worms than in mice or people, but their functions are quite similar. It has therefore been possible to use the simpler animals to gain knowledge that can be directly applied to humans. For example, the first netrin was discovered in a worm and shown to guide neurons around the worm's "nerve ring." Later, vertebrate netrins were found to guide axons around the mammalian spinal cord. Worm receptors for netrins were then found and proved invaluable in finding the corresponding, and again related, human receptors.

Once axons reach their targets, they form synapses, which permit electric signals in the axon to jump to the next cell, where they can either provoke or prevent the generation of a new signal. The regulation of this transmission at synapses, and the integration of inputs from the thousands of synapses each neuron receives, are responsible for the astounding information-processing capabilities of the brain. For processing to occur properly, the connections must be highly specific. Some specificity arises from the mechanisms that guide each axon to its proper target area. Additional molecules mediate "target recognition" whereby the axon chooses the proper neuron, and often the proper part of the target, once it arrives at its destination. Few of these molecules have been identified. There has been more success, however, in identifying the ways in which the synapse forms once the contact has been made. The tiny portion of the axon that contacts the dendrite becomes specialized for the release of neurotransmitters, and the tiny portion of the dendrite that receives the contact becomes specialized to receive and respond to the signal. Special molecules pass between the sending and receiving cell to ensure that the contact is formed properly.

Paring back

Following the period of growth, the network is pared back to create a more sturdy system. Only about one-half of the neurons generated during development survive to function in the adult. Entire populations of neurons are removed through internal suicide programs initiated in the cells. The programs are activated if a neuron loses its battle with other neurons to receive life-sustaining nutrients called trophic factors. These factors are produced in limited quantities by target tissues. Each type of trophic factor supports the survival of a distinct group of neurons. For example, nerve growth factor is important for sensory neuron survival. It has recently become clear that the internal suicide program is maintained into adulthood, and constantly held in check. Based on this idea, researchers have found that injuries and some neurodegenerative diseases kill neurons not directly by the damage they inflict, but rather by activating the death program. This discovery, and its implication that death need not inevitably follow insult, have led to new avenues for therapy.

Brain cells also form too many connections at first. For example, in primates, the projection from the two eyes to the brain initially overlaps, and then sorts out to separate territories devoted only to one or the other eye. Furthermore, in the young primate cerebral cortex, the connections between neurons are greater in number and twice as dense as an adult primate. Communication between neurons with chemical and electrical signals is necessary to weed out the connections. The connections that are active and generating electrical currents survive while those with little or no activity are lost.

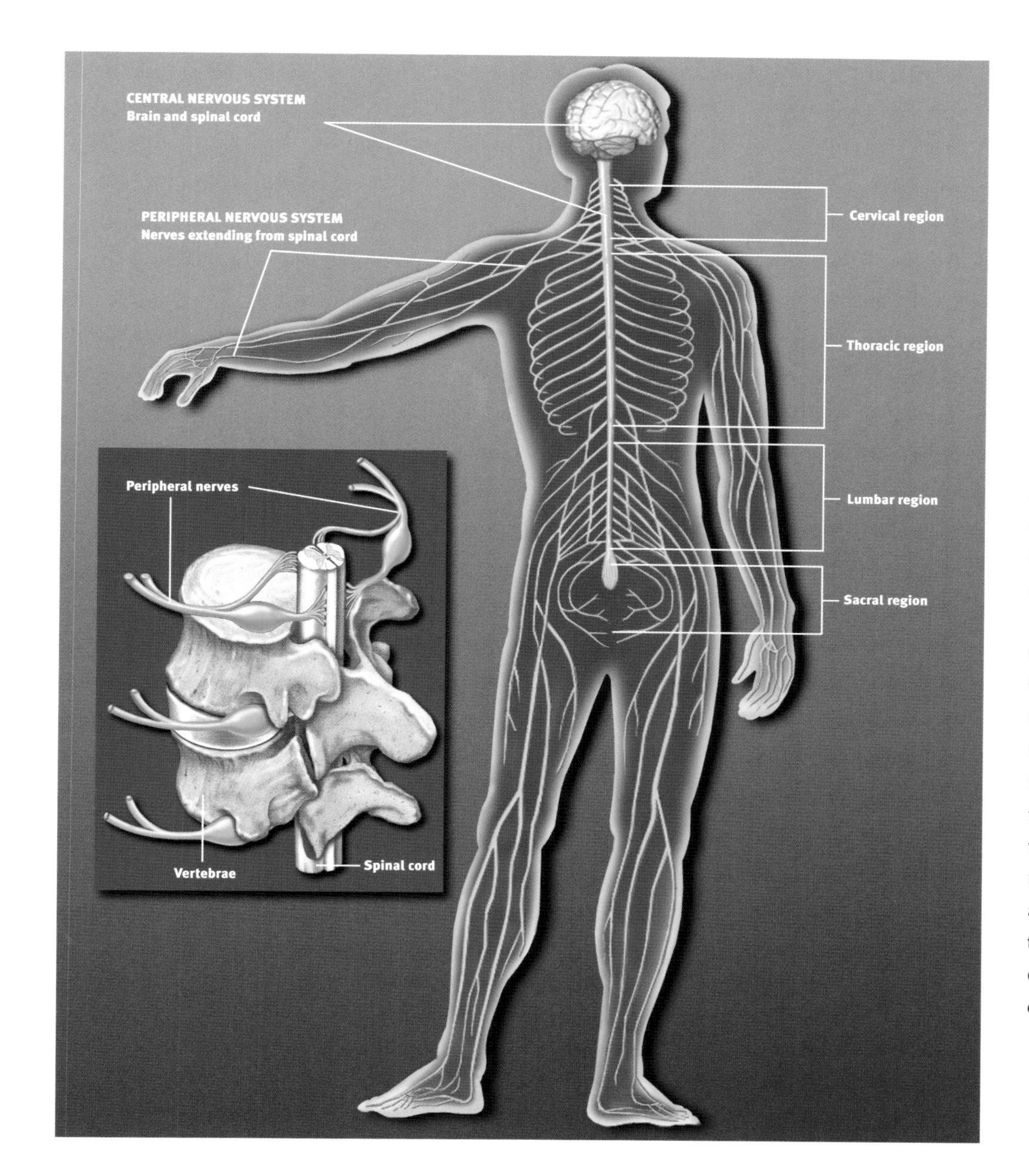

SPINAL CORD AND NERVES. The mature central nervous system (CNS) consists of the brain and spinal cord. The brain sends nerve signals to specific parts of the body through peripheral nerves, known as the peripheral nervous system (PNS). Peripheral nerves in the cervical region serve the neck and arms; those in the thoracic region serve the trunk; those in the lumbar region serve the legs; and those in the sacral region serve the bowels and bladder. The PNS consists of the somatic nervous system that connects voluntary skeletal muscles with cells specialized to respond to sensations, such as touch and pain. The autonomic nervous system is made of neurons connecting the CNS with internal organs. It is divided into the sympathetic nervous system, which mobilizes energy and resources during times of stress and arousal, and the parasympathetic nervous system, which conserves energy and resources during relaxed states.

Critical periods

The brain's refining and building of the network in mammals, including humans, continues after birth. An organism's interactions with its surroundings fine-tune connections.

Changes occur during critical periods. These are windows of time during development when the nervous system must obtain certain critical experiences, such as sensory, movement or emotional input, to develop properly. Following a critical period, connections become diminished in number and less subject to change, but the ones that remain are stronger, more reliable and more precise. Injury, sensory or social deprivation occurring at a certain stage of postnatal life may affect one aspect of development, while the same injury at a different period may affect another aspect. In one example, a monkey is raised from birth up to six months of age with one eyelid closed. As a result of diminished use, the animal permanently loses useful vision in that eye. This gives cellular meaning to the saying "use it or lose it." Loss of vision is caused by the actual loss of functional connections between that eye and neurons in the visual cortex. This finding has led to earlier and better treatment of the eye disorders congenital cataracts and "crossed-eyes" in children.

Research also shows that enriched environments can bolster brain development during postnatal life. For example, studies show that animals brought up in toy-filled surroundings have more branches on their neurons and more connections than isolated animals. In one recent study, scientists found enriched environments resulted in more neurons in a brain area involved in memory.

Scientists hope that new insights on development will lead to treatments for those with learning disabilities, brain damage and even neurodegenerative disorders or aging.

Sensation and perception

Vision. This wonderful sense allows us to image the world around us from the genius of Michelangelo's Sistine Chapel ceiling to mist-filled vistas of a mountain range. Vision is one of the most delicate and complicated of all the senses.

It also is the most studied. About one-fourth of the brain is involved in visual processing, more than for all other senses. More is known about vision than any other vertebrate sensory system, with most of the information derived from studies of monkeys and cats.

Vision begins with the *cornea*, which does about three-quarters of the focusing, and then the *lens*, which varies the focus. Both help produce a clear image of the visual world on the *retina*, the sheet of photoreceptors, which process vision, and neurons lining the back of the eye.

As in a camera, the image on the retina is reversed: objects to the right of center project images to the left part of the retina and vice versa. Objects above the center project to the lower part and vice versa. The shape of the lens is altered by the muscles of the *iris* so near or far objects can be brought into focus on the retina.

Visual receptors, about 125 million in each eye, are neurons specialized to turn light into electrical signals. They occur in two forms. *Rods* are most sensitive to dim light and do not convey the sense of color. *Cones* work in bright light and are responsible for acute detail, black and white and color vision. The human eye contains three types of cones that are sensitive to red, green and blue but in combination convey information about all visible colors.

Primates, including humans, have well-developed vision using two eyes. Visual signals pass from each eye along the million or so fibers of the optic nerve to the optic chiasma where some nerve fibers cross over, so both sides of the brain receive signals from both eyes. Consequently, the left halves of both retinae project to the left visual cortex and the right halves project to the right visual cortex.

The effect is that the left half of the scene you are watching registers in your right hemisphere. Conversely, the right half of the scene you are watching registers in your left hemisphere. A similar arrangement applies to movement and touch: each half of the cerebrum is responsible for the opposite half of the body.

Scientists know much about the way cells code visual information in the retina, *lateral geniculate nucleus*—an intermediate point between the retina and visual cortex—and visual cortex. These studies give us the best knowledge so far about how the brain analyzes and processes information.

The retina contains three stages of neurons. The first, the layer of rods and cones, sends its signals to the middle layer, which relays signals to the third layer. Nerve fibers from the third layer assemble to form the optic nerve. Each cell in the middle or third layer receives input from many cells in the previous layer. Any cell in the third layer thus receives signals—via the middle layer—from a cluster of many thousands of rods and cones that cover about one-square millimeter (the size of a thumb tack hole). This region is called the *receptive field* of the third-layer cell.

About 50 years ago, scientists discovered that the receptive field of such a cell is activated when light hits a tiny region in its receptive field center and is inhibited when light hits the part of the receptive field surrounding the center. If light covers the entire receptive field, the cell reacts only weakly and perhaps not at all.

Thus, the visual process begins with a comparison of the amount of light striking any small region of the retina and the amount of light around it. Located in the occipital lobe, the primary visual cortex—two millimeters thick (twice that of a dime) and densely packed with cells in many layers—receives messages from the lateral geniculate. In the middle layer, which receives input from the lateral geniculate, scientists found patterns of responsiveness similar to those observed in the retina and lateral geniculate cells. Cells above and below this layer responded differently. They preferred stimuli in the shape of bars or edges. Further studies showed that different cells preferred edges at particular angles, edges that moved or edges moving in a particular direction.

Although the process is not yet completely understood,

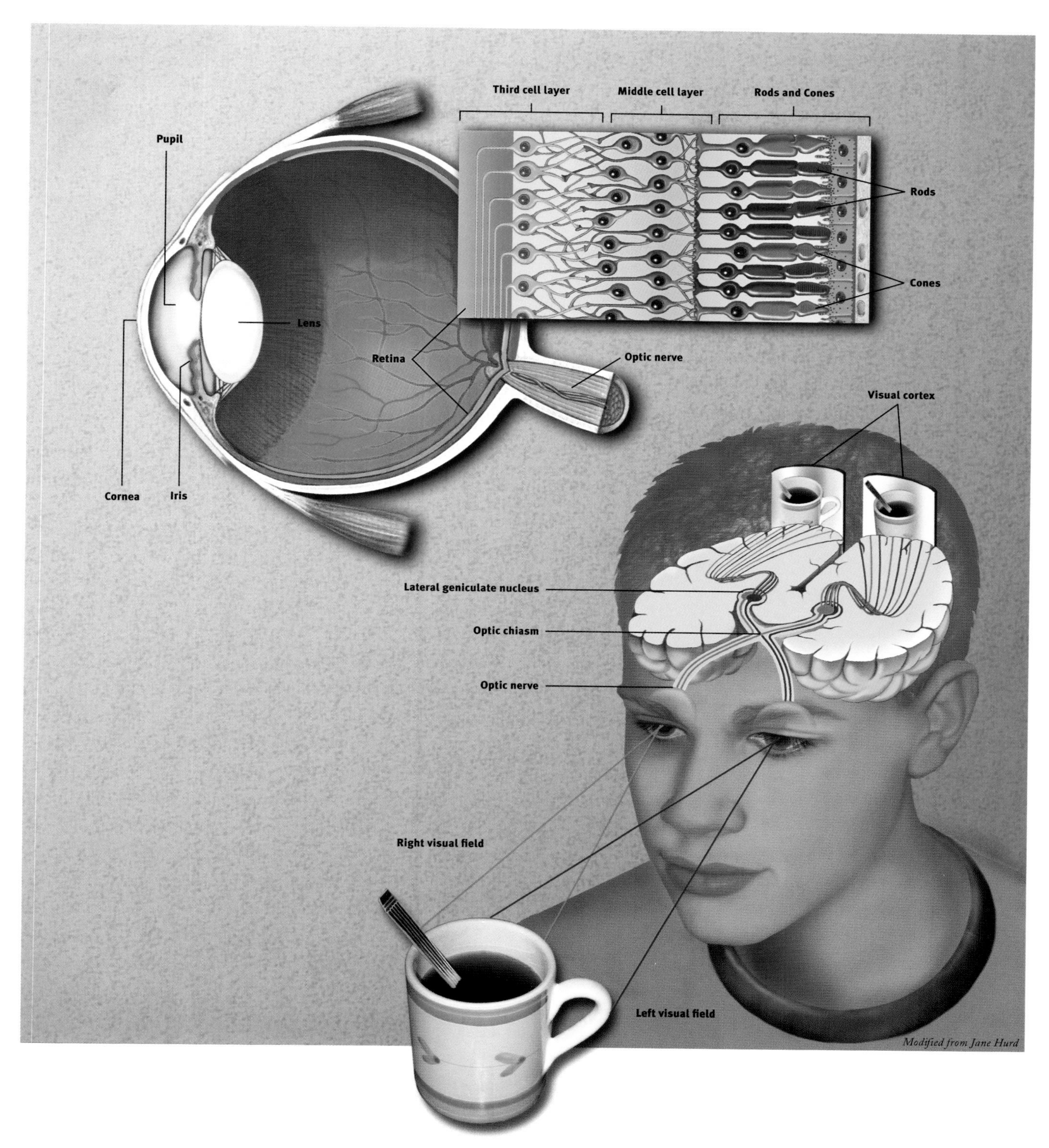

VISION. The cornea and lens help produce a clear image of the visual world on the retina, the sheet of photoreceptors and neurons lining the back of the eye. As in a camera, the image on the retina is reversed: objects to the right of center project images to the left part of the retina and vice versa. The eye's 125 million visual receptors—composed of rods and cones—turn light into electrical signals. Rods are most sensitive to dim light and do not convey the sense of color; cones work in bright light and are responsible for acute detail, black and white and color vision. The human eye contains three types of cones that are sensitive to red, green and blue but, in combination, convey information about all visible colors. Rods and cones connect with a middle cell layer and third cell layer (see inset, above). Light passes through these two layers before reaching the rods and cones. The two layers then receive signals from rods and cones before transmitting the signals onto the optic nerve, optic chiasm, lateral geniculate nucleus and, finally, the visual cortex.

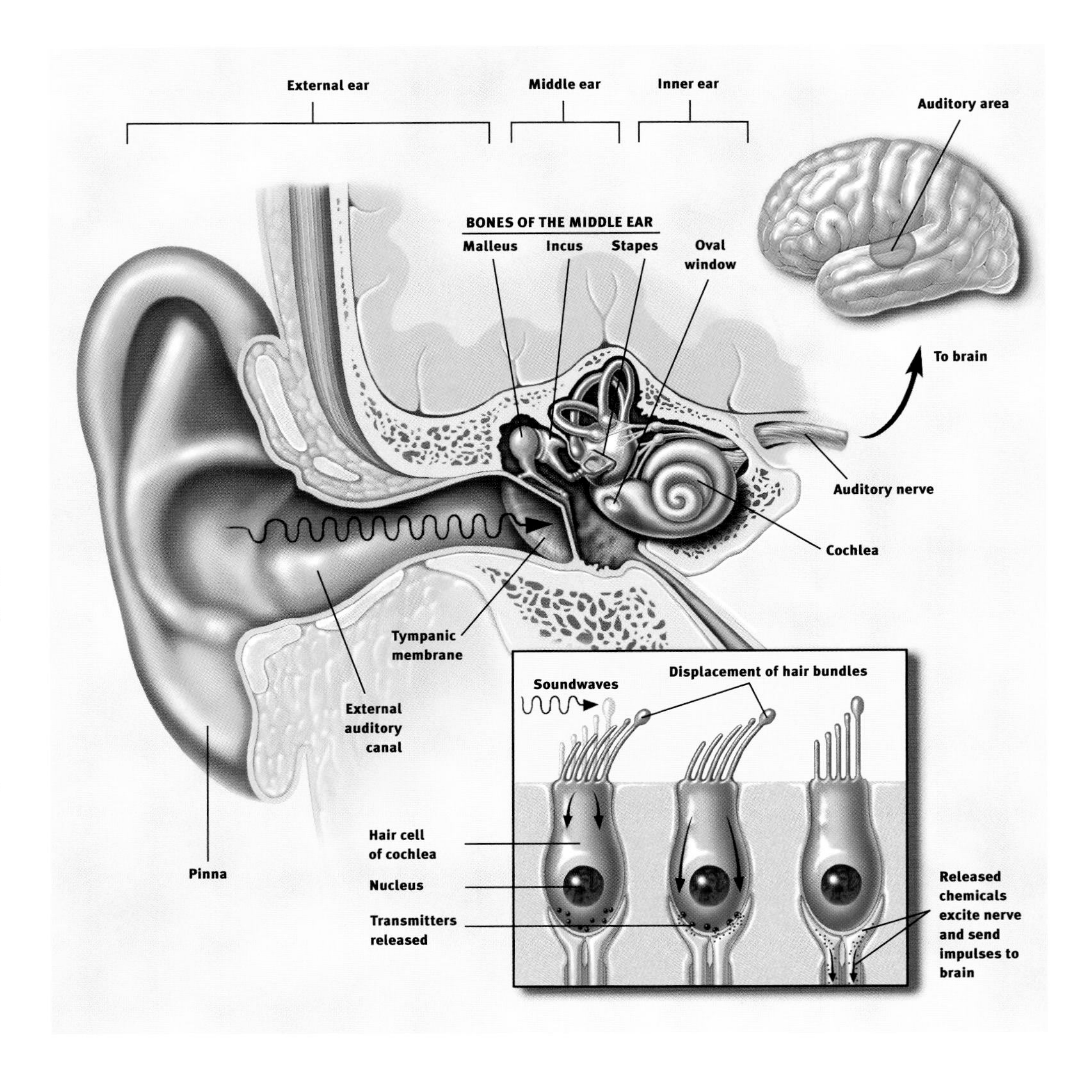

HEARING. From the chirping of crickets to the roar of a rocket engine, almost all of the thousands of single tones processed by the human ear are heard by a mechanism known as air conduction. In this process, sound waves are first funneled through the external ear—the pinna and the external auditory canal—to the middle ear—the tympanic membrane (eardrum) that vibrates at different speeds. The malleus (hammer), which is attached to the tympanic membrane, transmits the vibrations to the incus (anvil). The vibrations are then passed onto the stapes (stirrup) and oval window that, in turn, pass them onto the inner ear. In the inner ear, the fluid-filled spiral passage of the cochlea contains cells with microscopic, hairlike projections that respond to the vibrations produced by sound. The hair cells, in turn, excite the 28,000 fibers of the auditory nerve that end in the medulla in the brain. Auditory information flows via the thalamus to the temporal gyrus, the part of the cerebral cortex involved in receiving and perceiving sound.

recent findings suggest that visual signals are fed into at least three separate processing systems. One system appears to process information about shape; a second, color; and a third, movement, location and spatial organization. These findings of separate processing systems come from monkey anatomical and physiological data. They are verified by human psychological studies showing that the perception of movement, depth, perspective, the relative size of objects, the relative movement of objects and shading and gradations in texture all depend primarily on contrasts in light intensity rather than in color.

Why movement and depth perception should be carried by only one processing system may be explained by a school of thought called Gestalt psychology. Perception requires various elements to be organized so that related ones are grouped together. This stems from the brain's ability to group the parts of an image together and also to separate images from one another and from their individual backgrounds.

How do all these systems produce the solid images you see? By extracting biologically relevant information at each stage and associating firing patterns with past experience.

Vision studies also have led to better treatment for visual disorders. Information from research in cats and monkeys has improved the therapy for *strabismus*, or *squint*, a term for "cross-eye" or wall-eye. Children with strabismus initially have good vision in each eye. But because they cannot fuse the images in the two eyes, they tend to favor using one eye and often lose useful vision in the other eye.

Vision can be restored but only during infancy or early childhood. Beyond the age of six or so, the blindness becomes permanent. But until a few decades ago, ophthalmologists waited until children reached the age of four before operating to align the eyes, or prescribe exercises or an eye patch. Now strabismus is corrected very early in life—before age four—when normal vision can still be restored.

Hearing

Often considered the most important sense for humans, hearing allows us to communicate with each other by receiving sounds and interpreting speech. It also gives us information vital to survival. For example, the sound of an oncoming train tells us to stay clear of the railroad track.

Like the visual system, our hearing system distinguishes several qualities in the signal it detects. However, our hearing system does not blend different sounds, as the visual system does when two different wavelengths of light are mixed to produce color. We can follow the separate melodic lines of several instruments as we listen to an orchestra or rock band.

From the chirping of crickets to the roar of a rocket engine, most of the sounds processed by the ear are heard by a mechanism known as *air conduction*. In this process, sound waves are first funneled through the externally visible part of the ear, the pinna (or external ear) and the *external auditory canal* to the *tympanic membrane* (eardrum) that vibrates at different speeds. The *malleus* (hammer), which is attached to the tympanic membrane, transmits the vibrations to the *incus* (anvil). This structure passes them onto the *stapes* (stirrup) which delivers them, through the oval window, to the *inner ear*.

The fluid-filled spiral passages of each cochlea contain 16,000 hair cells whose microscopic, hairlike projections respond to the vibrations produced by sound. The hair cells, in turn, excite the 28,000 fibers of the auditory nerve that terminate in the medulla of the brain. Auditory information flows via the thalamus to the *temporal gyrus*, the part of the cerebral cortex involved in receiving and perceiving sound.

The brain's analysis of auditory information follows a pattern similar to that of the visual system. Adjacent neurons respond to tones of similar frequency. Some neurons respond to only a small range of frequencies, others react to a wide range; some react only to the beginning of a sound, others only respond to the end.

Speech sounds, however, may be processed differently than others. Our auditory system processes all the signals that it receives in the same way until they reach the primary auditory cortex in the temporal lobe of the brain. When speech sound is perceived, the neural signal is funneled to the left hemisphere for processing in language centers.

Taste and smell are two separate senses with their own sets of receptor organs, but they act together to distinguish an enormous number of different flavors.

Taste and smell

Although different, the two sensory experiences of taste and smell are intimately entwined. They are separate senses with their own receptor organs. However, these two senses act together to allow us to distinguish thousands of different flavors. Alone, taste is a relatively focused sense concerned with distinguishing among sweet, salty, sour and bitter. The interaction between taste and smell explains why loss of the sense of smell apparently causes a serious reduction in the overall taste experience, which we call flavor.

Tastes are detected by *taste buds*, special structures of which every human has some 5,000. Taste buds are embedded within *papillae*, or protuberances, located mainly on the tongue, with others found in the back of the mouth and on the palate. Taste substances stimulate hairs projecting from the sensory cells. Each taste bud consists of 50 to 100 sensory cells that respond to salts, acidity, sweet substances and bitter compounds. Some researchers add a fifth category named *umami*, for the taste of monosodium glutamate and related substances.

Taste signals in the sensory cells are transferred by synapses to the ends of nerve fibers, which send impulses along cranial nerves to taste centers in the brain. From here, the impulses are relayed to other brain stem centers responsible for the basic responses of acceptance or rejection of the tastes, and to the thalamus and on to the cerebral cortex for conscious perception of taste.

Specialized smell receptor cells are located in a small patch of mucus membrane lining the roof of the nose. Axons of these sensory cells pass through perforations in the overlying bone and enter two elongated *olfactory bulbs* lying on top of the bone. The portion of the sensory cell that is exposed to odors possesses hair-like cilia. These cilia contain the receptor sites that are stimulated by odors carried by airborne molecules. The odor molecules dissolve in the mucus lining in order to stimulate receptor molecules in the cilia to start the smell response. An odor molecule acts on many receptors to different degrees. Similarly, a receptor interacts with many different odor molecules to different degrees.

Axons of the cells pass through perforations in the overlying bone and enter two elongated olfactory bulbs lying on top of the bone. The pattern of activity set up in the receptor cells is projected to the olfactory bulb, where it forms a spatial image of the odor. Impulses created by this stimulation pass to smell centers, to give rise to conscious perceptions of odor in the frontal lobe and emotional responses in the limbic system of the brain.

SMELL AND TASTE. Specialized receptors for smell are located in a patch of mucous membrane lining the roof of the nose. Each cell has several fine hairlike cilia containing receptor proteins, which are stimulated by odor molecules in the air, and a long fiber (axon), which passes through perforations in the overlying bone to enter the olfactory bulb. Stimulated cells give rise to impulses in the fibers, which set up patterns in the olfactory bulb that are relayed to the brain's frontal lobe to give rise to smell perception, and to the limbic system to elicit emotional responses. Tastes are detected by special structures, taste buds, of which every human has some 10,000. Taste buds are embedded within papillae (protuberances) mainly on the tongue, with a few located in the back of the mouth and on the palate. Each taste bud consists of about 100 receptors that respond to the four types of stimuli—sweet, salty, sour and bitter—from which all tastes are formed. A substance is tasted when chemicals in foods dissolve in saliva, enter the pores on the tongue and come in contact with taste buds. Here they stimulate hairs projecting from the receptor cells and cause signals to be sent from the cells, via synapses, to cranial nerves and taste centers in the brain.

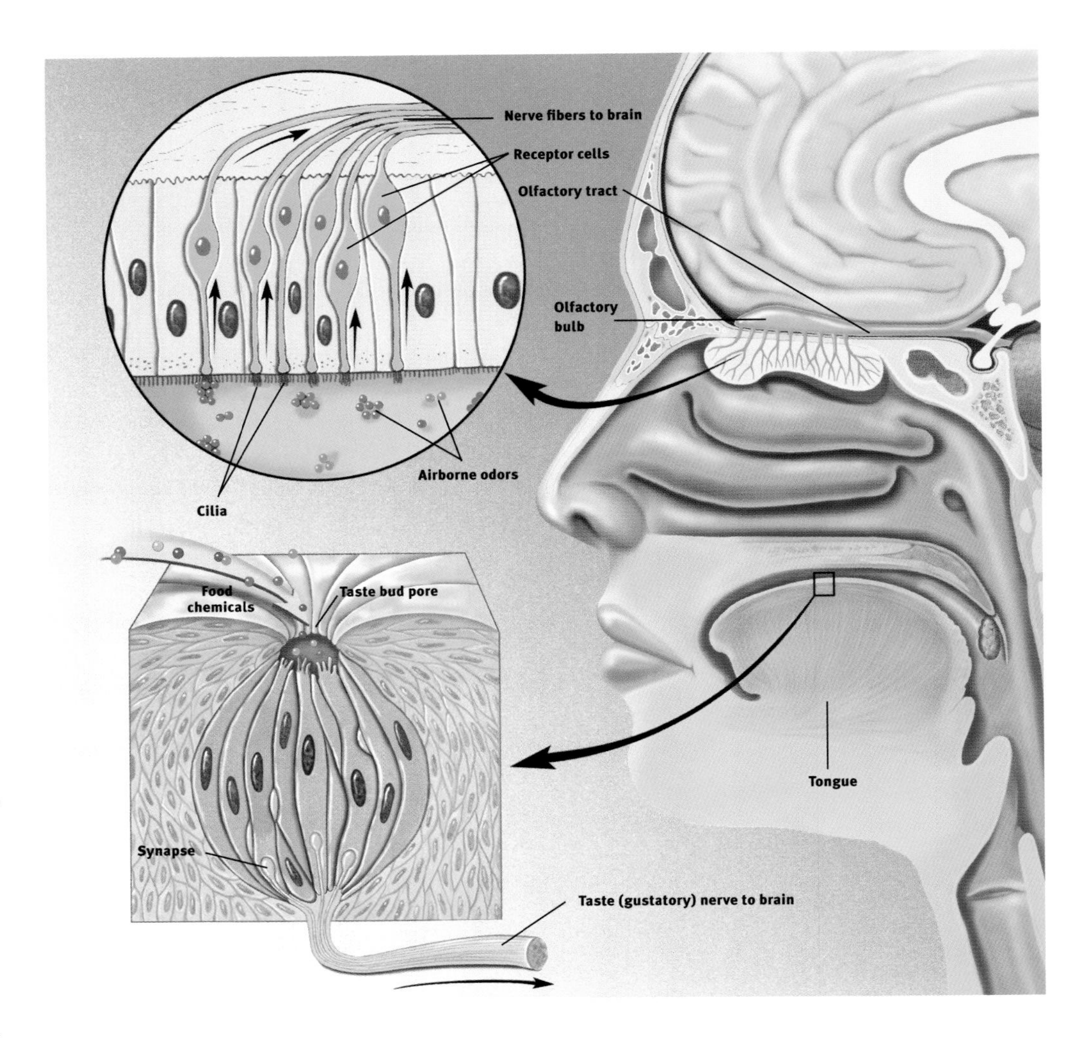

Touch and pain

Touch is the sense by which we determine the characteristics of objects: size, shape and texture. We do this through touch receptors in the skin. In hairy skin areas, some receptors consist of webs of sensory nerve cell endings wrapped around the hair bulbs. They are remarkably sensitive, being triggered when the hairs are moved. Other receptors are more common in non-hairy areas, such as lips and fingertips, and consist of nerve cell endings that may be free or surrounded by bulb-like structures.

Signals from touch receptors pass via sensory nerves to the spinal cord, then to the thalamus and sensory cortex. The transmission of this information is highly topographic, meaning that the body is represented in an orderly fashion at different levels of the nervous system. Larger areas of the cortex are devoted to sensations from the hands and lips; much smaller cortical regions represent less sensitive parts of the body.

Different parts of the body vary in their sensitivity to touch discrimination and painful stimuli according to the number and distribution of receptors. The cornea is several hundred times more sensitive to painful stimuli than are the soles of the feet. The fingertips are good at touch discrimination but the chest and back are less sensitive.

Until recently, pain was thought to be a simple message by which neurons sent electrical impulses from the site of injury directly to the brain.

Recent studies show that the process is more complicated. Nerve impulses from sites of injury that persist for hours, days or longer lead to changes in the nervous system that result in an amplification and increased duration of the pain. These changes involve dozens of chemical messengers and receptors.

At the point of injury, *nociceptors*, special receptors, respond to tissue-damaging stimuli. Injury results in the release of numerous chemicals at the site of damage and inflammation. One such chemical, *prostaglandin*, enhances the sensitivity of receptors to tissue damage and ultimately can result in more intense pain sensations. It also contributes to the clinical condition in which innocuous stimuli can produce pain (such as in sunburned skin) because the threshold of the nociceptor is significantly reduced.

Pain messages are transmitted to the spinal cord via small myelinated fibers and C fibers—very small unmyelinated fibers. Myelin is a covering around nerve fibers that helps them send their messages more rapidly.

In the *ascending system*, the impulses are relayed from the spinal cord to several brain structures, including the thalamus and cerebral cortex, which are involved in the process by which "pain" messages become conscious experience.

Pain messages can also be suppressed by a system of neurons that originate within the gray matter in the brainstem of the midbrain. This *descending system* sends messages to the dorsal horn of the spinal cord where it suppresses the transmission of pain signals to the higher brain centers. Some of these descending systems use naturally occurring chemicals similar to opioids. The three major families of opioids—enkephalins, endorphins and dynorphins—identified in the brain originate from three precursor proteins coded by three different genes. They act at multiple opioid receptors in the brain and spinal cord. This knowledge has led to new treatments for pain: Opiate-like drugs injected into the space above the spinal cord provide long-lasting pain relief.

Scientists are now using modern tools for imaging brain structures in humans to determine the role of the higher centers of the brain in pain experience and how signals in these structures change with long-lasting pain.

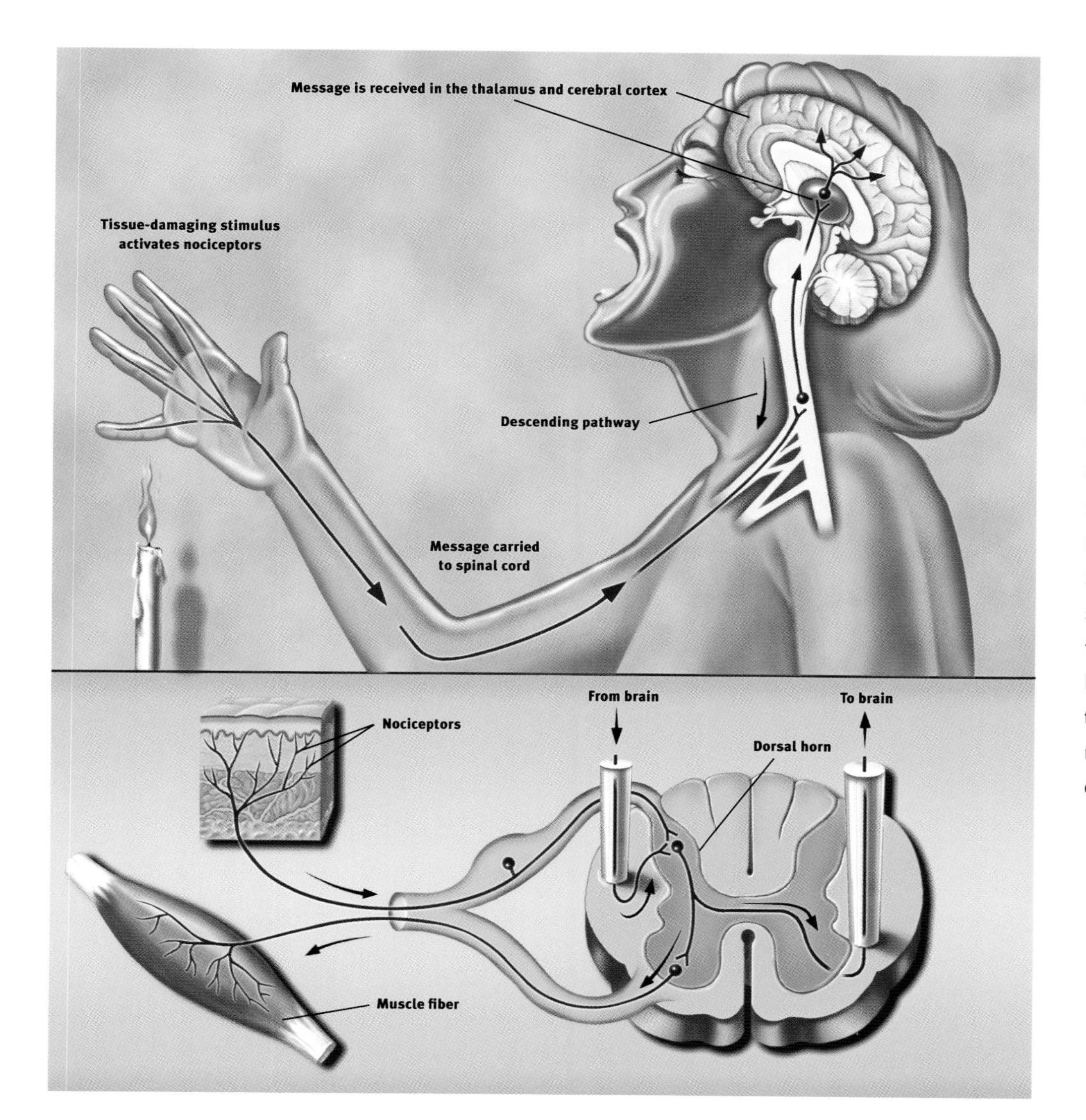

PAIN. Messages about tissue damage are picked up by receptors and transmitted to the spinal cord via small, myelinated fibers and very small unmyelinated fibers. From the spinal cord, the impulses are carried to the brainstem, thalamus and cerebral cortex and ultimately perceived as pain. These messages can be suppressed by a system of neurons that originates in the gray matter of the midbrain. This descending pathway sends messages to the spinal cord where it suppresses the transmission of tissue damage signals to the higher brain centers. Some of these descending pathways use naturally occurring, opiate-like chemicals called endorphins.

Learning and memory

The conscious memory of a patient known as H.M. is limited almost entirely to events that occurred years before his surgery, which removed part of the medial temporal lobe of his brain to relieve epilepsy. H.M. can remember recent events for only a few minutes. Talk with him awhile and then leave the room. When you return, he has no recollection of ever having seen you before.

The medial temporal lobe, which includes the hippocampus and adjacent brain areas, seems to play a role in converting memory from a short-term to a long-term, permanent form. The fact that H.M. retains memories for events that are remote to his surgery is evidence that the medial temporal region is not the site of permanent storage but that it plays a role in the formation of new memories. Other patients like H.M. have also been described.

Additional evidence comes from patients undergoing *electroconvulsive therapy* (ECT) for depression. ECT is thought to temporarily disrupt the function of the hippocampus and related structures. These patients typically suffer difficulty with new learning and have amnesia for events that occurred during the several years before treatment. Memory of earlier events is unimpaired. As time passes after treatment, much of the lost part of memory becomes available once again.

The hippocampus and the medial temporal region are connected with widespread areas of the cerebral cortex, especially the vast regions responsible for thinking and language. Whereas the medial temporal region is important for forming and organizing memory, cortical areas are important for the long-term storage of knowledge about facts and events and for how these are used in everyday situations.

Working memory, a type of transient memory that enables us to retain what someone has said just long enough to reply, depends in part on the prefrontal cortex. Researchers discovered that certain neurons in this area are influenced by neurons releasing dopamine and other neurons releasing glutamate.

While much is unknown about learning and memory, scientists can recognize certain pieces of the process. For example, the brain appears to process different kinds of information in separate ways and then store it differently. *Procedural knowledge*, the knowledge of how to do something, is expressed in skilled behavior and learned habits. *Declarative knowledge* provides an explicit, consciously accessible record of individual previous experiences and a sense of familiarity about those experiences. Declarative knowledge requires processing in the medial temporal region and parts of the thalamus, while procedural knowledge requires processing by the basal ganglia. Other kinds of memory depend on the amygdala (emotional aspects of memory) and the cerebellum (motor learning where precise timing is involved).

An important factor that influences what is stored and how strongly it is stored is whether the action is followed by rewarding or punishing consequences. This is an important principle in determining what behaviors an organism will learn and remember. The amygdala appears to play an important role in these memory events.

How exactly does memory occur? After years of study, there is much support for the idea that memory involves a persistent change in the relationship between neurons. In animal studies, scientists found that this occurs through biochemical events in the short term that affect the strength of the relevant synapses. The stability of long-term memory is conferred by structural modifications within neurons that change the strength and number of synapses. For example, researchers can correlate specific chemical and structural changes in the relevant cells with several simple forms of behavioral change exhibited by the sea slug *Aplysia*.

Another important model for the study of memory is the phenomenon of *long-term potentiation* (LTP), a long-lasting increase in the strength of a synaptic response following stimulation. LTP occurs prominently in the hippocampus, as well as in other brain areas. Studies of rats suggest LTP occurs by changes in synaptic strength at contacts involving NMDA receptors. It is now possible to study LTP and learning in genetically modified mice that have abnormalities of specific genes. Abnormal gene expression can be limited to particular brain areas and time periods, such as during learning.

Scientists believe that no single brain center stores memory. It most likely is stored in the same, distributed collection

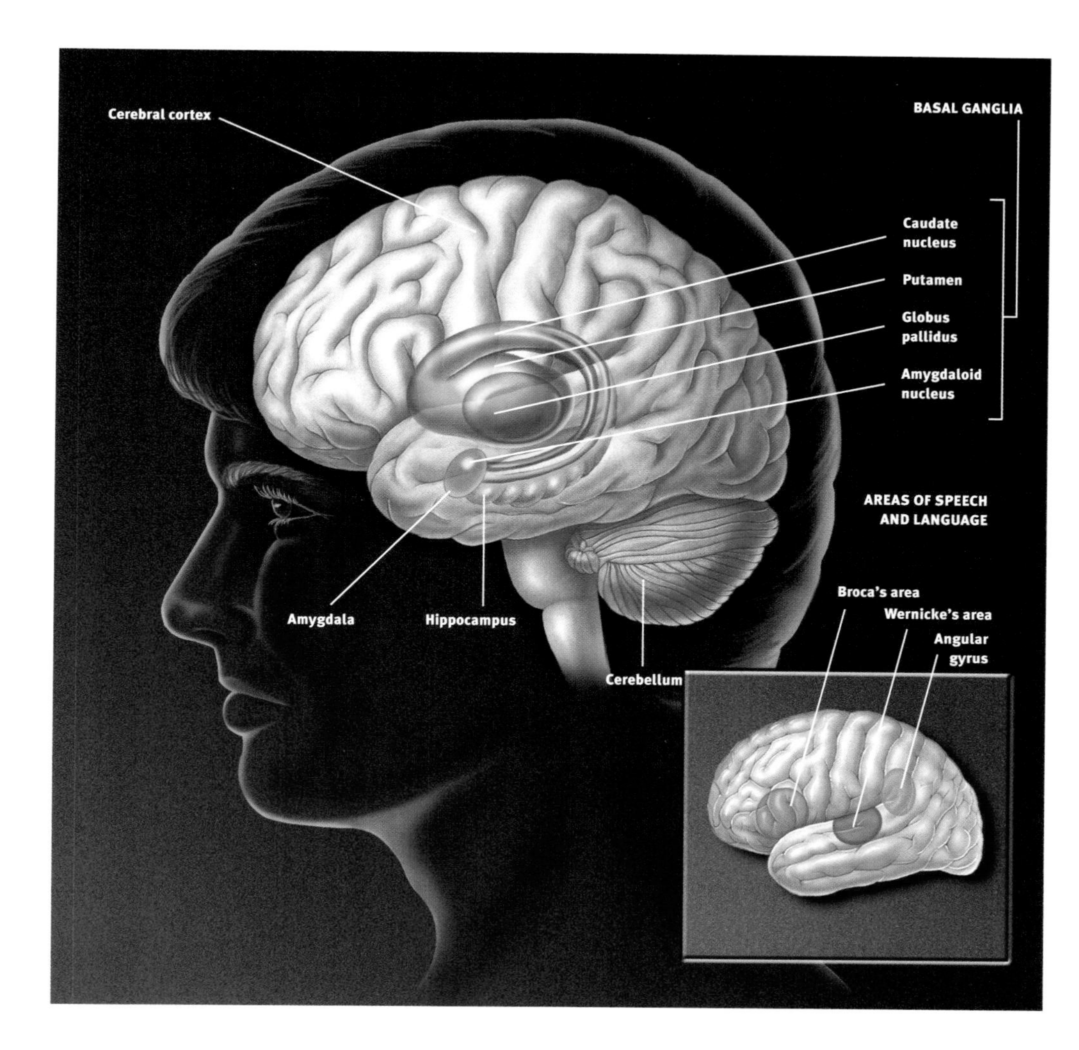

LEARNING AND MEMORY, SPEECH AND LANGUAGE. Structures believed to be important for various kinds of learning and memory include the cerebral cortex, amygdala, hippocampus, cerebellum and basal ganglia. Areas of the left hemisphere (inset) are known to be active in speech and language. The form and meaning of an utterance is believed to arise in Wernicke's area and then Broca's area, which is related to vocalization. Wernicke's area is also important for language comprehension.

of cortical processing systems involved in the perception, processing and analysis of the material being learned. In short, each part of the brain most likely contributes differently to permanent memory storage.

One of the most prominent intellectual activities dependent on memory is language. While the neural basis of language is not fully understood, scientists have learned much about this feature of the brain from studies of patients who have lost speech and language abilities due to stroke, and from behavioral and functional neuroimaging studies of normal people.

A prominent and influential model, based on studies of these patients, proposes that the underlying structure of speech comprehension arises in Wernicke's area, a portion of the left hemisphere of the brain. This temporal lobe region is connected with Broca's area in the frontal lobe where a program for vocal expression is created. This program is then transmitted to a nearby area of the motor cortex that activates the mouth, tongue and larynx.

This same model proposes that, when we read a word, the information is transmitted from the primary visual cortex to the angular gyrus where the message is somehow matched with the sounds of the words when spoken. The auditory form of the word is then processed for comprehension in Wernicke's area as if the word had been heard. Writing in response to an oral instruction requires information to be passed along the same pathways in the opposite direction—from the auditory cortex to Wernicke's area to the angular gyrus. This model accounts for much of the data from patients, and is the most widely used model for clinical diagnosis and prognosis. However, some refinements to this model may be necessary due to both recent studies with patients and functional neuroimaging studies in normal people.

For example, using an imaging technique called *positron emission tomography* (PET), scientists have demonstrated that some reading tasks performed by normal people activated neither Wernicke's area nor the angular gyrus. These results suggest that there is a direct reading route that does not involve speech sound recoding of the visual stimulus before the processing of either meaning or speaking. Other studies with patients also have indicated that it is likely that familiar words need not be recoded into sound before they can be understood.

Although the understanding of how language is implemented in the brain is far from complete, there are now several techniques that may be used to gain important insights.

Movement

From the stands, we marvel at the perfectly placed serves of professional tennis players and lightning-fast double plays executed by big league infielders. But in fact, every one of us in our daily lives performs highly skilled movements, such as walking upright, speaking and writing, that are no less remarkable. A finely tuned and highly complex central nervous system controls the action of hundreds of muscles in accomplishing these everyday marvels.

In order to understand how the nervous system performs this trick, we have to start with muscles. Most muscles attach to points on the skeleton that cross one or more joints. Activation of a given muscle, the *agonist*, can open or close the joints that it spans or act to stiffen them, depending on the forces acting on those joints from the environment or other muscles that oppose the agonist, the *antagonists*. Relatively few muscles act on soft tissue. Examples include the muscles that move the eyes and tongue, and the muscles that control facial expression.

A muscle is made up of thousands of individual muscle fibers, each of which is controlled by one *alpha motor neuron* in either the brain or spinal cord. On the other hand, a single alpha neuron can control hundreds of muscle fibers, forming a *motor unit*. These motor neurons are the critical link between the brain and muscles. When these neurons die, a person is no longer able to move.

The simplest movements are reflexes—fixed muscle responses to particular stimuli. Studies show sensory stretch receptors—called *muscle spindles*, which include small, specialized muscle fibers and are located in most muscles—send information about muscles directly to alpha motor neurons.

Sudden muscle stretch (such as when a doctor taps a muscle tendon to test your reflexes) sends a barrage of impulses into the spinal cord along the muscle spindle sensory fibers. This, in turn, activates motor neurons in the stretched muscle, causing a contraction which is called the stretch reflex. The same sensory stimulus causes inactivation, or inhibition, in the motor neurons of the antagonist muscles through connecting neurons, called *inhibitory neurons*, within the spinal cord.

The sensitivity of the muscle spindle organs is controlled by the brain through a separate set of *gamma motor neurons* that control the specialized spindle muscle fibers and allow the brain to fine-tune the system for different movement tasks. Other muscle sense organs signal muscle force that affects motor neurons through separate sets of spinal neurons. We now know that this complex system responds differently for tasks that require precise control of position (holding a full teacup), as opposed to those that require rapid, strong movement (throwing a ball). You can experience such changes in motor strategy when you compare walking down an illuminated staircase with the same task done in the dark.

Another useful reflex is the *flexion withdrawal* that occurs if your bare foot encounters a sharp object. Your leg is immediately lifted from the source of potential injury (flexion) but the opposite leg responds with increased extension in order to maintain your balance. The latter event is called the *crossed extension reflex*. These responses occur very rapidly and without your attention because they are built into systems of neurons located within the spinal cord itself.

It seems likely that the same systems of spinal neurons also participate in controlling the alternating action of the legs during normal walking. In fact, the basic patterns of muscle activation that produce coordinated walking can be generated in four-footed animals within the spinal cord itself. It seems likely that these spinal mechanisms, which evolved in primitive vertebrates, are probably still present in the human spinal cord.

The most complex movements that we perform, including voluntary ones that require conscious planning, involve control of the spinal mechanisms by the brain. Scientists are only beginning to understand the complex interactions that take place between different brain regions during voluntary movements, mostly through careful experiments on animals. One important area is the *motor cortex*, which exerts powerful control of the spinal cord neurons and has direct control of some motor neurons in monkeys and humans. Some neurons in the motor cortex appear to specify the coordinated action of many muscles, so as to produce organized movement of the limb to a particular place in space.

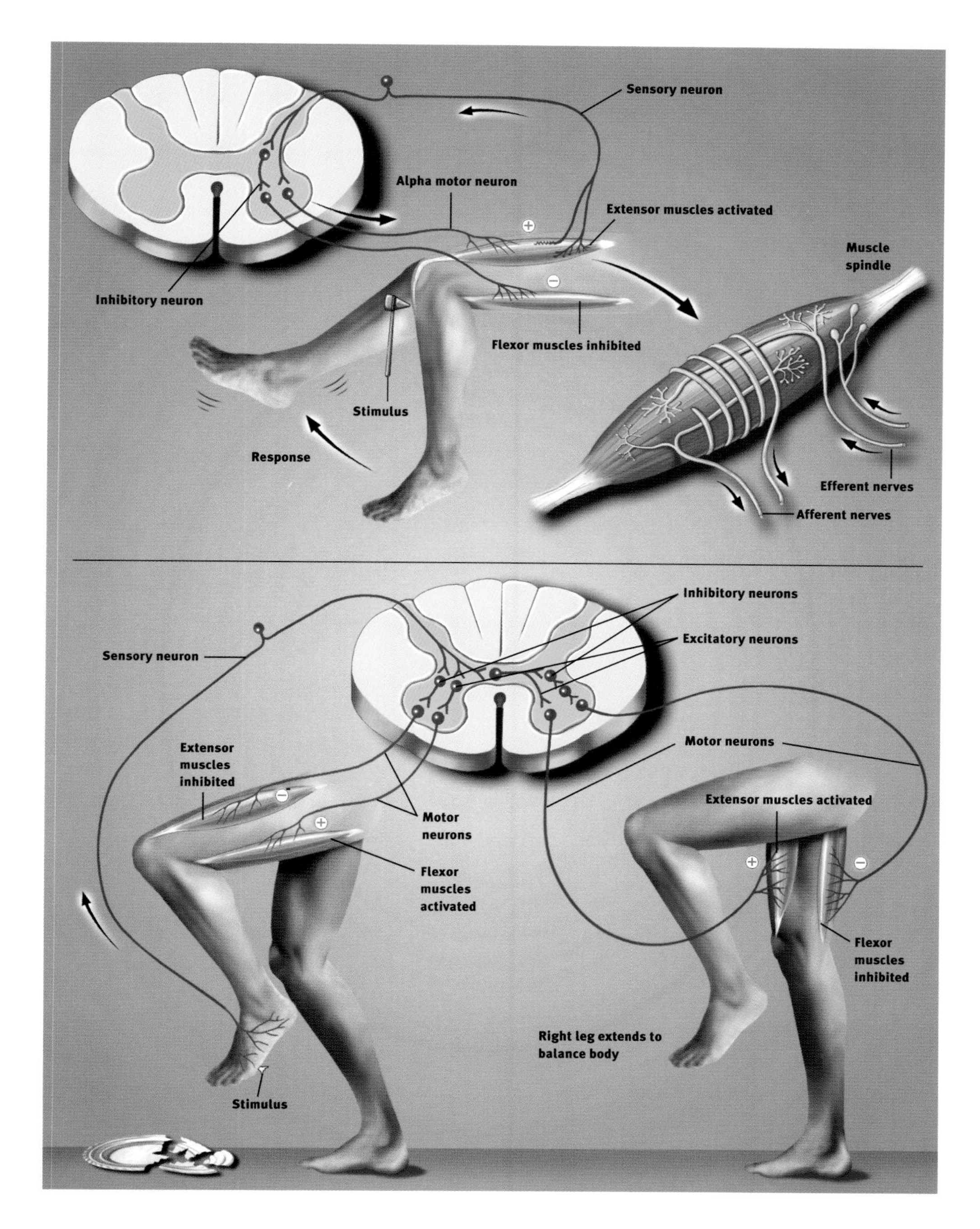

MOVEMENT. The stretch reflex (above) occurs when a doctor taps a muscle tendon to test your reflexes. This sends a barrage of impulses into the spinal cord along muscle spindle sensory fibers and activates motor neurons to the stretched muscle to cause contraction (stretch reflex). The same sensory stimulus causes inactivation, or inhibition, of the motor neurons to the antagonist muscles through connection neurons, called inhibitory neurons, within the spinal cord. Afferent nerves carry messages from sense organs to the spinal cord; efferent nerves carry motor commands from the spinal cord to muscles. Flexion withdrawal (below) can occur when your bare foot encounters a sharp object. Your leg is immediately lifted (flexion) from the source of potential injury, but the opposite leg responds with increased extension in order to maintain your balance. The latter event is called the crossed extension reflex. These responses occur very rapidly and without your attention because they are built into systems of neurons located within the spinal cord itself.

In addition to the motor cortex, movement control also involves the interaction of many other brain regions, including the basal ganglia and thalamus, the cerebellum and a large number of neuron groups located within the midbrain and brainstem—regions that connect cerebral hemispheres with the spinal cord.

Scientists know that the basal ganglia and thalamus have widespread connections with sensory and motor areas of the cerebral cortex. Loss of regulation of the basal ganglia by dopamine depletion can cause serious movement disorders, such as Parkinson's disease. Loss of dopamine neurons in the substantia nigra on the midbrain, which connects with the basal ganglia, is a major factor in Parkinson's.

The cerebellum is critically involved in the control of all skilled movements. Loss of cerebellar function leads to poor coordination of muscle control and disorders of balance. The cerebellum receives direct and powerful sensory information from the muscle receptors, and the sense organs of the inner ear, which signal head position and movement, as well as signals from the cerebral cortex. It apparently acts to integrate all this information to ensure smooth coordination of muscle action, enabling us to perform skilled movements more or less automatically. There is evidence that, as we learn to walk, speak or play a musical instrument, the necessary detailed control information is stored within the cerebellum where it can be called upon by commands from the cerebral cortex.

Sleep

Sleep remains one of the great mysteries of modern neuroscience. We spend nearly one-third of our lives asleep, but the function of sleep still is not known. Fortunately, over the last few years researchers have made great headway in understanding some of the brain circuitry that controls wake-sleep states.

Scientists now recognize that sleep consists of several different stages; that the choreography of a night's sleep involves the interplay of these stages, a process that depends upon a complex switching mechanism; and that the sleep stages are accompanied by daily rhythms in bodily hormones, body temperature and other functions.

Sleep disorders are among the nation's most common health problems, affecting up to 70 million people, most of whom are undiagnosed and untreated. These disorders are one of the least recognized sources of disease, disability and even death, costing an estimated $100 billion annually in lost productivity, medical bills and industrial accidents. Research holds the promise for devising new treatments to allow millions of people to get a good night's sleep.

The stuff of sleep

Sleep appears to be a passive and restful time when the brain is less active. In fact, this state actually involves a highly active and well-scripted interplay of brain circuits to produce the stages of sleeping.

The stages of sleep were discovered in the 1950s in experiments examining the human brain waves or electroencephalogram (EEG) during sleep. Researchers also measured movements of the eyes and the limbs during sleep. They found that over the course of the first hour or so of sleep each night, the brain progresses through a series of stages during which the brain waves progressively slow down. The period of *slow wave sleep* is accompanied by relaxation of the muscles and the eyes. Heart rate, blood pressure and body temperature all fall. If awakened at this time, most people recall only a feeling or image, not an active dream.

SLEEP PATTERNS. During a night of sleep, the brain waves of a young adult recorded by the electroencephalogram (EEG) gradually slow down and become larger as the individual passes into deeper stages of slow wave sleep. After about an hour, the brain re-emerges through the same series of stages, and there is usually a brief period of REM sleep (on dark areas of graph), during which the EEG is similar to wakefulness. The body is completely relaxed, the person is deeply unresponsive and usually is dreaming. The cycle repeats over the course of the night, with more REM sleep, and less time spent in the deeper stages of slow wave sleep as the night progresses.

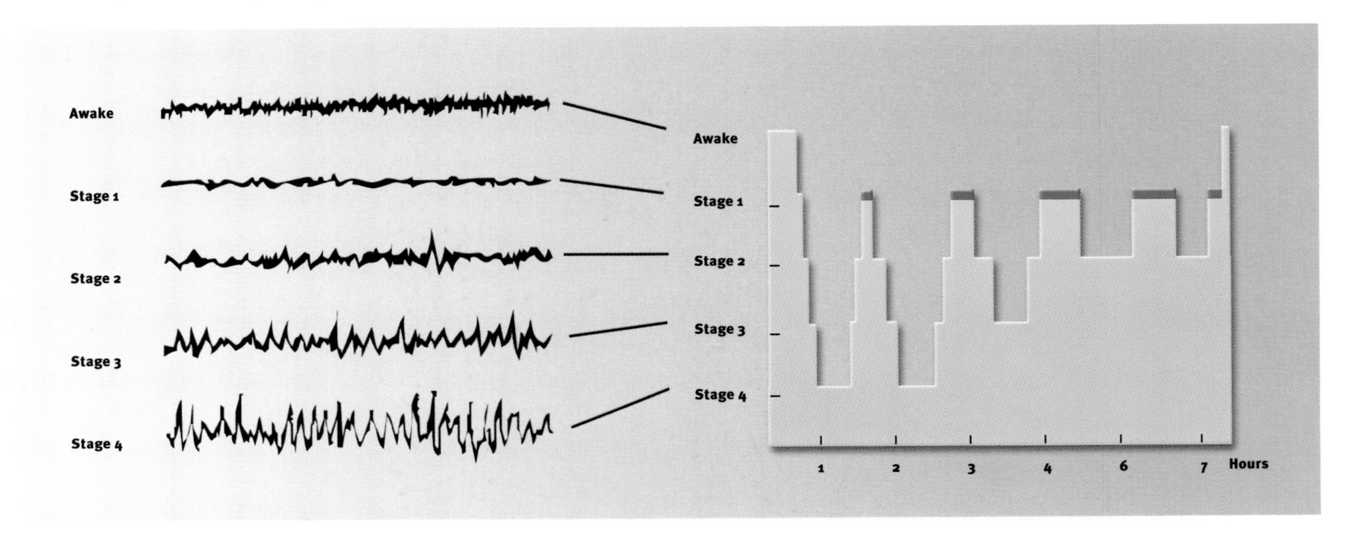

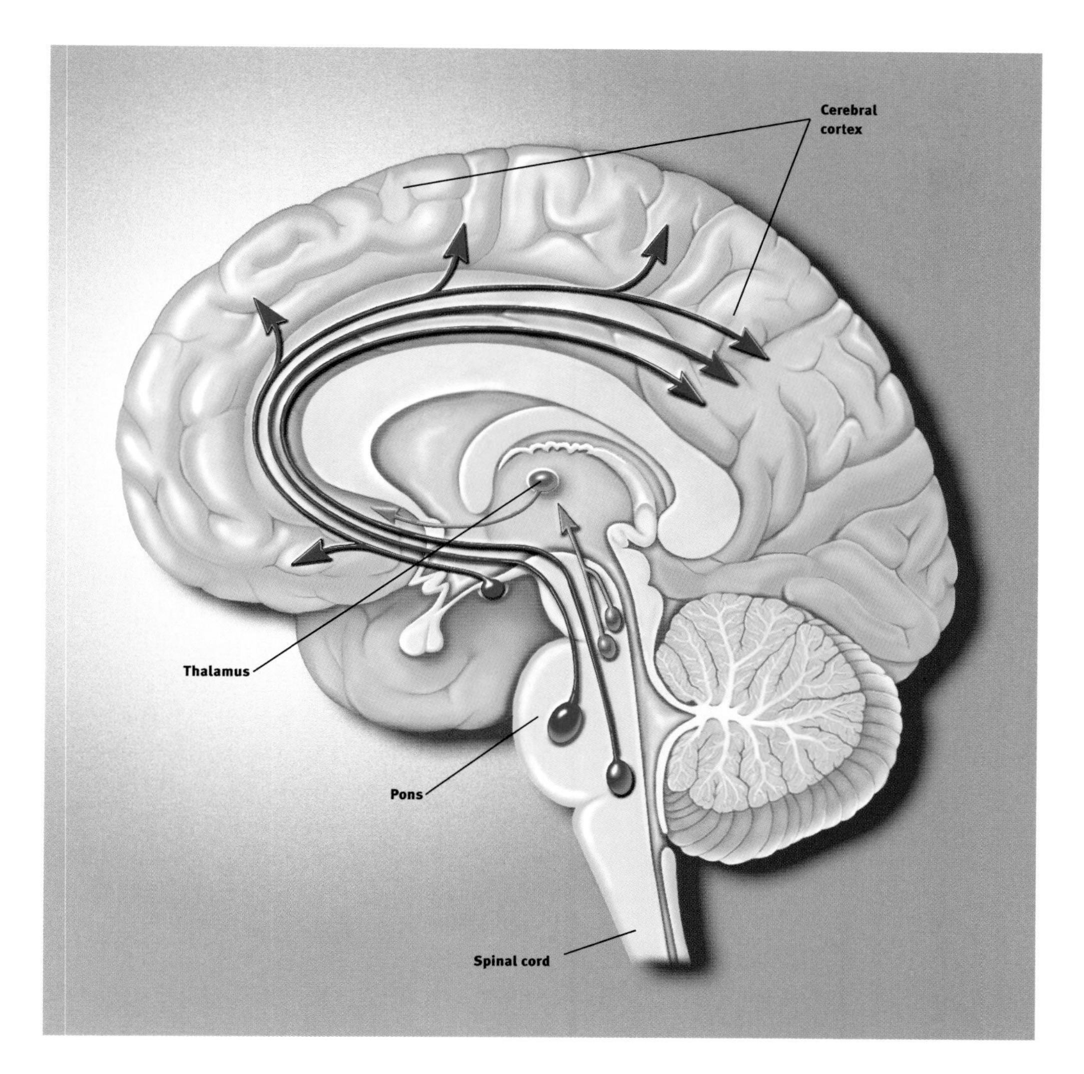

THE WAKING AND SLEEPING BRAIN. Wakefulness is maintained by activity in two systems of brainstem neurons. Nerve cells that make the neurotransmitter acetylcholine stimulate the thalamus, which activates the cerebral cortex (red pathway). Full wakefulness also requires cortical activation by other neurons that make monoamine neurotransmitters such as norepinephrine, serotonin and histamine (blue pathway). During slow wave sleep, when the brain becomes less active, neuron activity in both pathways slows down. During rapid eye movement sleep, in which dreaming occurs, the neurons using acetylcholine fire rapidly, producing a dreaming state, but the monoamine cells stop firing altogether.

Over the next half hour or so, the brain emerges from the deep slow wave sleep as the EEG waves become progressively faster. Similar to during waking, rapid eye movements emerge, but the body's muscles become almost completely paralyzed (only the muscles that allow breathing remain active). This state is often called *rapid eye movement* (REM) sleep. During REM sleep, there is active dreaming. Heart rate, blood pressure and body temperature become much more variable. Men often have erections during this stage of sleep. The first REM period usually lasts ten to 15 minutes.

Over the course of the night, these alternative cycles of slow wave and REM sleep alternate, with the slow wave sleep becoming less deep, and the REM periods more prolonged, until waking occurs.

Over the course of a lifetime, the pattern of sleep cycles changes. Infants sleep up to 18 hours per day, and they spend much more time in deep slow wave sleep. As children mature, they spend less time asleep, and less time in deep slow wave sleep. Older adults may sleep only six to seven hours per night, often complain of early wakening that they cannot avoid, and spend very little time in slow wave sleep.

Sleep disorders

The most common sleep disorder, and the one most people are familiar with, is *insomnia*. Some people have difficulty falling asleep initially, but other people fall asleep, and then awaken part way through the night, and cannot fall asleep again. Although there are a variety of short-acting sedatives and sedating antidepressant drugs available to help, none of these produces a truly natural and restful sleep state because they tend to suppress the deeper stages of slow wave sleep.

Excessive daytime sleepiness may have many causes. The most common are disorders that disrupt sleep and result in inadequate amounts of sleep, particularly the deeper stages. These are usually diagnosed in the sleep laboratory. Here, the EEG, eye movements and muscle tone are monitored electrically as the individual sleeps. In addition, the heart, breathing, and oxygen content of the blood can be monitored.

Obstructive sleep apnea causes the airway muscles in the throat to collapse as sleep deepens. This prevents breathing, which causes arousal, and prevents the sufferer from entering the deeper stages of slow wave sleep. This condition can also cause high blood pressure and may increase the risk of heart attack. There is also an increased risk of daytime accident, especially automobile accidents, which may prevent driving. Treatment is complex and may include a variety of attempts to reduce airway collapse during sleep. While simple things like losing weight, avoiding alcohol and sedating drugs prior to sleep, and avoiding sleeping on one's back can sometimes help, most people with sleep apnea require positive airway pressure to keep the airway open. This can be provided by fitting a small mask over the nose that provides an air stream under pressure during sleep. In some cases, surgery is needed to correct the airway anatomy.

Periodic limb movements of sleep are intermittent jerks of the legs or arms, which occur as the individual enters slow wave sleep, and can cause arousal from sleep. Other people have episodes in which their muscles fail to be paralyzed during REM sleep, and they act out their dreams. This *REM behavior disorder* can also be very disruptive to a normal nights' sleep. Both disorders are more common in people with Parkinson's disease, and both can be treated with drugs that treat Parkinson's, or with an anti-epileptic drug called clonazepam.

Narcolepsy is a relatively uncommon condition (one case per 2,500 people) in which the switching mechanism for REM sleep does not work properly. Narcoleptics have sleep attacks during the day, in which they suddenly fall asleep. This is socially disruptive, as well as dangerous, for example, if they are driving. They tend to enter REM sleep very quickly as well, and may even enter a dreaming state while still awake, a condition known as *hypnagogic hallucinations.* They also have attacks during which they lose muscle tone, similar to what occurs during REM sleep, but while they are awake. Often, this occurs while they are falling asleep or just waking up, but attacks of paralysis known as *cataplexy* can be triggered by an emotional experience or even hearing a funny joke.

Recently, insights into the mechanism of narcolepsy have given major insights into the processes that control these mysterious transitions between waking, slow wave and REM sleep states.

How is sleep regulated?

During wakefulness, the brain is kept in an alert state by the interactions of two major systems of nerve cells. Nerve cells in the upper part of the pons and in the midbrain, which make acetylcholine as their neurotransmitter, send inputs to the thalamus, to activate it. When the thalamus is activated, it in turn activates the cerebral cortex, and produces a waking EEG pattern. However, that is not all there is to wakefulness. As during REM sleep, the cholinergic nerve cells and the thalamus and cortex are in a condition similar to wakefulness, but the brain is in REM sleep, and is not very responsive to external stimuli.

The difference is supplied by three sets of nerve cells in the upper part of the brainstem: nerve cells in the locus coeruleus that contain the neurotransmitter norepinephrine; in the dorsal and median raphe groups that contain serotonin; and in the tuberomammillary cell group that contains histamine. These monoamine neurons fire most rapidly during wakefulness, but they slow down during slow wave sleep, and they stop during REM sleep.

The brainstem cell groups that control arousal are in turn regulated by two groups of nerve cells in the hypothalamus, part of the brain that controls basic body cycles. One group of nerve cells, in the ventrolateral preoptic nucleus, contain inhibitory neurotransmitters, galanin and GABA. When the ventrolateral preoptic neurons fire, they are thought to turn off the arousal systems, causing sleep. Damage to the ventrolateral preoptic nucleus produces irreversible insomnia.

A second group of nerve cells in the lateral hypothalamus act as an activating switch. They contain the neurotransmitters orexin and dynorphin, which provide an excitatory signal to the arousal system, particularly to the monoamine neurons. In experiments in which the gene for the neurotransmitter orexin was experimentally removed in mice, the animals became narcoleptic. Similarly, in two dog strains with naturally occurring narcolepsy, an abnormality was discovered in the gene for the type 2 orexin receptor. Recent studies show that in humans with narcolepsy, the orexin levels in the brain and spinal fluid are abnormally low. Thus, orexin appears to play a critical role in activating the monoamine system, and preventing abnormal transitions, particularly into REM sleep.

Two main signals control this circuitry. First, there is homeostasis, or the body's need to seek a natural equilibrium. There is an intrinsic need for a certain amount of sleep each day. The mechanism for accumulating sleep need is not yet clear. Some people think that a chemical called adenosine may accumulate in the brain during prolonged wakefulness, and that it may drive sleep homeostasis. Interestingly, the drug caffeine, which is widely used to prevent sleepiness, acts as an adenosine blocker, to prevent its effects.

If an individual does not get enough sleep, the sleep debt progressively accumulates, and leads to a degradation of mental function. When the opportunity comes to sleep again, the individual will sleep much more, to "repay" the debt, and the slow wave sleep debt is usually "paid off" first.

The other major influence on sleep cycles is the body's circadian clock, the suprachiasmatic nucleus. This small group of nerve cells in the hypothalamus contains clock genes, which go through a biochemical cycle of almost exactly 24 hours, setting the pace for daily cycles of activity, sleep, hormones and other bodily functions. The suprachiasmatic nucleus also receives an input directly from the retina, and the clock can be reset by light, so that it remains linked to the outside world's day-night cycle. The suprachiasmatic nucleus provides a signal to the ventrolateral preoptic nucleus and probably the orexin neurons.

Stress

The urge to act in the presence of stress has been with us since our ancient ancestors. In response to impending danger, muscles are primed, attention is focused and nerves are readied for action—fight or flight. But in today's corporation-dominated world, this response to stress is simply inappropriate and may be a contributor to heart disease, accidents and aging.

Indeed, nearly two-thirds of ailments seen in doctors' offices are commonly thought to be stress-induced or related to stress in some way. Surveys indicate that 60 percent of Americans feel they are under a great deal of stress at least once a week. Costs due to stress from absenteeism, medical expenses and lost productivity are estimated at $300 billion annually.

Only recently admitted into the medical vocabulary, stress is difficult to define because its effects vary with each individual. Dr. Hans Selye, a founder of stress research, called it "the rate of wear and tear in the body." Other specialists now define stress as any external stimulus that threatens homeostasis—the normal equilibrium of body function. Among the most powerful stressors are psychological and psychosocial stressors that exist between members of the same species. Lack or loss of control is a particularly important feature of severe psychological stress that can have physiological consequences.

During the last six decades, researchers using animals found that stress both helps and harms the body. When confronted with a crucial challenge, properly controlled stress responses can provide the extra strength and energy needed to cope. Moreover, the acute physiological response to stress protects the body and brain and helps to re-establish or maintain homeostasis. But stress that continues for prolonged periods of time can repeatedly elevate the physiological stress responses or fail to shut them off when not needed. When this occurs, these same physiological mechanisims can badly upset the body's biochemical balance and accelerate disease.

Scientists also believe that the individual variation in responding to stress is somewhat dependent on a person's perception of external events. This perception ultimately shapes his or her internal physiological response. Thus, by controlling your perception of events, you can do much to avoid the harmful consequences of stress.

The immediate response

A stressful situation activates three major communication systems in the brain that regulate bodily functions. Scientists have come to understand these complex systems through experiments primarily with rats, mice and nonhuman primates, such as monkeys. Scientists then verified the action of these systems in humans.

The first of these systems is the *voluntary nervous system*, which sends messages to muscles so that we may respond to sensory information. For example, the sight of a growling bear on a trail in Yellowstone National Park prompts you to run as quickly as possible.

The second communication system is the *autonomic nervous system*. It combines the sympathetic or emergency branch, which gets us going in emergencies, and the *parasympathetic* or calming branch, which keeps the body's maintenance systems, such as digestion, in order and calms the body's responses to the emergency branch.

Each of these systems has a specific task. The emergency branch causes arteries supplying blood to the muscles to relax in order to deliver more blood, allowing greater capacity to act. At the same time, the emergency system reduces blood flow to the skin, kidney and digestive tract and increases blood flow to the muscles. In contrast, the calming branch helps to regulate bodily functions and soothe the body, preventing it from remaining too long in a state of mobilization. Remaining mobilized and left unchecked, these body functions could lead to disease. Some actions of the calming branch appear to reduce the harmful effects of the emergency branch's response to stress.

The brain's third major communication process is the *neuroendocrine system*, which also maintains the body's internal functioning. Various "stress hormones" travel through the blood and stimulate the release of other hormones, which affect bodily processes, such as metabolic rate and sexual functions.

Major stress hormones are *epinephrine* (also known as adrenaline) and *cortisol*. When the body is exposed to stressors,

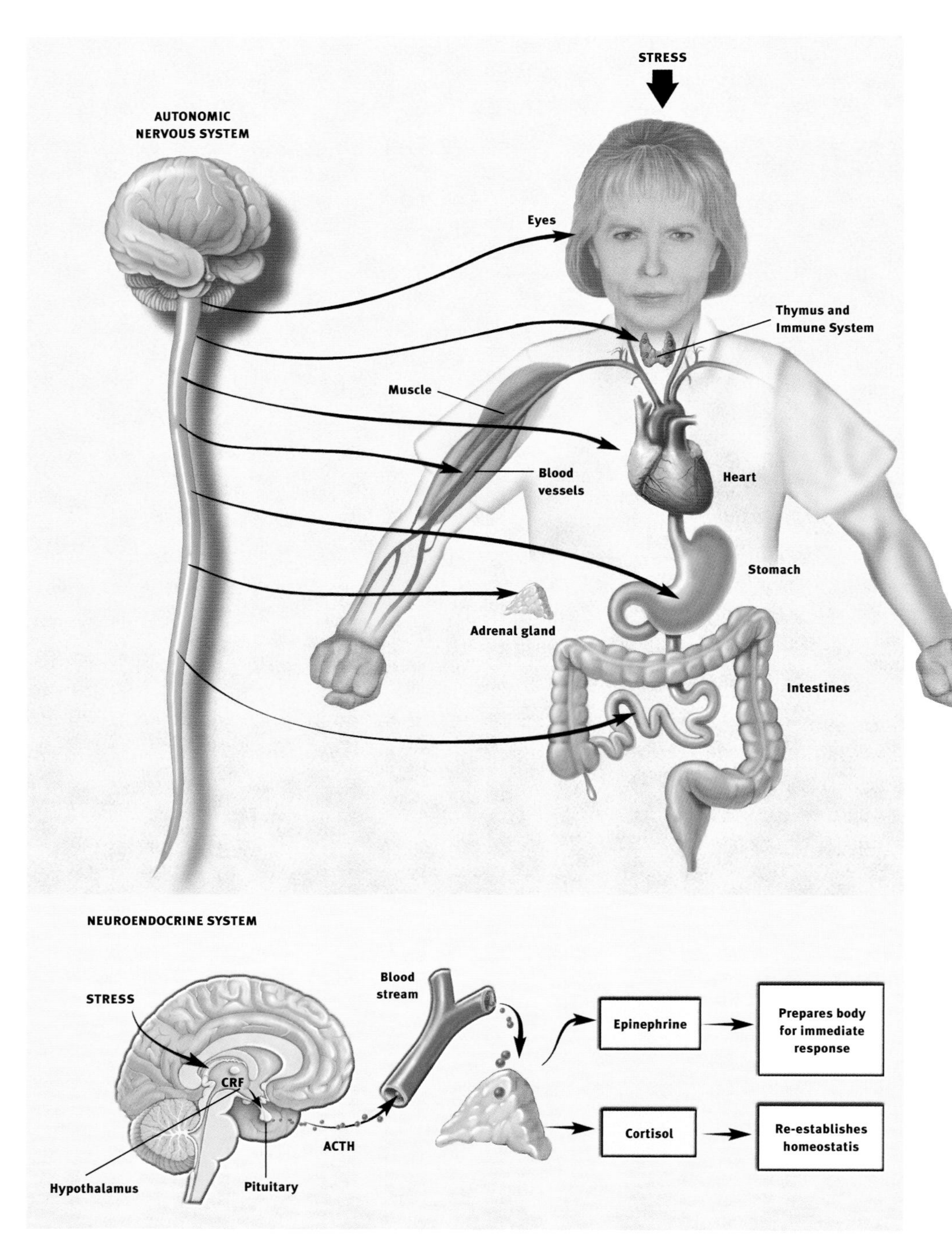

THE STRESS REACTION. When stress occurs, the sympathetic nervous system is triggered. Norepinephrine is released by nerves, and epinephrine is secreted by the adrenal glands. By activating receptors in blood vessels and other structures, these substances ready the heart and working muscles for action.

In the parasympathetic nervous system, acetylcholine is released, producing calming effects. The digestive tract is stimulated to digest a meal, the heart rate slows and the pupils of the eye become smaller.

The neuroendocrine system also maintains the body's normal internal functioning. Corticotrophin-releasing factor (CRF), a peptide formed by chains of amino acids, is released from the hypothalamus, a collection of cells at the base of the brain that acts as a control center for the neuroendocrine system. CRF travels to the pituitary gland where it triggers the release of adrenocorticotropic hormone (ACTH). ACTH travels in the blood to the adrenal glands where it stimulates the release of cortisol.

epinephrine is quickly released into the bloodstream to put the body into a general state of arousal and enable it to cope with a challenge.

The secretion by the adrenal glands of cortisol—known as a *glucocorticoid* because it affects the metabolism of glucose, a source of energy—starts about five minutes later. Some of its actions help to mediate the stress-response, while some of its other, slower ones, counteract the primary response to stress and help re-establish homeostasis. Over the short run, cortisol mobilizes energy and delivers it to muscles for the body's response. With prolonged exposure, cortisol enhances feeding and helps the body recover from energy mobilization.

Acute stress also increases activity of the immune system and helps protect the body from disease pathogens. The two major stress hormones, cortisol and adrenaline, facilitate the movement of immune cells from the bloodstream and storage organs such as the spleen into tissue where they are needed to defend against an infection.

Glucocorticoids also affect food intake during the sleep-wake cycle. Cortisol levels peak in the body in the early morning hours just before waking. This hormone acts as a wake-up signal and helps to turn on appetite and physical activity. This effect of glucocorticoids may help to explain disorders, such as jet-lag, which result when the light-dark cycle is altered by jet travel over long distances, causing the body's biological clock to reset itself more slowly. Until that clock is reset, cortisol secretion and hunger, as well as sleepiness and wakefulness, occur at inappropriate times of day in the new location.

Glucocorticoids do more than help the body respond to stress. In fact, they are an integral part of daily life and the adaptation to environmental change. The adrenal glands help protect us from stress and are essential for survival.

Chronic stress

When glucocorticoids or epinephrine are secreted in response to the prolonged psychological stress commonly encountered by humans, the results are not ideal. Normally, bodily systems gear up under stress, and hormones are released to enhance muscular activity and restore homeostasis. If you are not fighting or fleeing—but standing frustrated in a supermarket check-out line or sitting in a traffic jam—you are not engaging in muscular exercise. Yet these systems continue to be stimulated.

Stressful experiences have a direct effect on heart rate and blood pressure. When stressors are chronic and psychological, the effect can be extremely harmful and result in an increased risk for heart attack.

Overexposure to cortisol also can lead to weakened muscles and the suppression of major bodily systems. Elevated epinephrine production increases blood pressure. Together, elevated cortisol and epinephrine can contribute to chronic hypertension (high blood pressure), abdominal obesity and atherosclerosis (hardening of the arteries).

Scientists have identified a variety of stress-related disorders, including colitis, high blood pressure, clogged arteries, impotency and loss of sex drive in males, irregular menstrual cycles in females, adult-onset diabetes and possibly cancer. Aging rats show impairment of neuronal function in the hippocampus—an area of the brain important for learning, memory and emotion—as a result of cortisol secretion throughout their lifetimes. Overexposure to glucocorticoids also increases the number of neurons damaged by stroke. Moreover, prolonged exposure before or immediately after birth can cause a decrease in the normal number of brain neurons and smaller brain size.

The *immune system*, which receives messages from the nervous system, also is sensitive to many of the circulating hormones of the body, including stress hormones. Moderate to high levels of glucocorticoids act to suppress immune function, although acute elevations of stress hormones actually facilitate immune function.

Although acute stress-induced immunoenhancement can be protective against disease pathogens, the glucocorticoid-induced immunosuppression can also be beneficial. It reduces inflammation and counteracts allergic reactions and *autoimmune responses* that occur when the body's defenses turn against body tissue. Synthetic glucocorticoids like hydrocortisone and prednisone are used all the time to decrease inflammation and autoimmunity. But glucocorticoids may be harmful in the case of increased tumor growth associated with stress in experiments on animals—an area of intense research yet to yield any final conclusions.

One important determinant of the immune system's resistance or susceptibility to disease may be a person's sense of control as opposed to a feeling of helplessness. This phenomenon may help explain large individual variations in response to disease. Scientists are trying to identify how the perception of control or helplessness influences physiological processes that regulate immune function.

The *cardiovascular system* receives many messages from the autonomic nervous system, and stressful experiences have an immediate and direct effect on heart rate and blood pressure. In the short run, these changes help in response to stressors. But when stressors are chronic and psychological, the effect can be harmful and result in accelerated atherosclerosis and increased risk for heart attack. Research supports the idea that people holding jobs that carry high demands and low control, such as telephone operators, waiters and cashiers, have higher rates of heart disease than people who can dictate the pace and style of their working lives.

Behavioral type affects a person's susceptibility to heart attack. People at greatest risk are hostile, irritated by trivial things and exhibit signs of struggle against time and other challenges.

Researchers found that two groups of men—one with high hostility scores and the other with low hostility scores—exhibited similar increases in blood pressure and blood muscle flow when performing a lab test. This confirmed that hostility scores do not predict the biological response to simple mental tasks.

Then the researchers added harassment to the test by leading the subjects to believe that their performances were being unfairly criticized. Men with high hostility scores showed much larger increases in muscle blood flow and blood pressure, and slower recovery than those with low hostility scores. Scientists found that harassed men with high hostility scores have larger increases in levels of stress hormones. Thus, if you are a hostile person, learning to reduce or avoid anger could be important to avoid cardiovascular damage.

Aging

Pablo Picasso, Georgia O'Keefe and Grandma Moses, artists. Louise Nevelson, sculptor. Albert Einstein, physicist. Giuseppe Verdi, musician. Robert Frost, poet. Each of these great minds worked differently, but they all shared an outstanding trait: They were creative and productive in old age. They defied the popular notion that aging always leads to a pronounced decline and loss of cognitive ability.

Neuroscientists now believe that the brain can remain relatively healthy and fully functioning as it ages, and that diseases are the causes of the most severe decline in memory, intelligence, verbal fluency and other tasks. Researchers are investigating the normal changes that occur over time and the effect that these changes have on reasoning and other intellectual activities.

It appears that the effects of age on brain function vary widely. The vast majority of people get only a bit forgetful in old age, particularly in forming memories of recent events. For example, once you reach your 70s, you may start to forget names or phone numbers, or respond more slowly to conflicting information. This is not disease. However, other individuals develop *senile dementia*, the progressive and severe impairment in mental function that interferes with daily living. The senile dementias include Alzheimer's and cerebrovascular diseases and affect about one percent of people younger than age 65, with the incidence increasing to nearly 50 percent in those older than 85. In a small, third group, that includes the Picassos, Nevelsons and others, mental functioning seems unaffected by age. Many people do well throughout life and continue to do well even when they are old. The oldest human, Jeanne Calment, was considered to have all her wits during her 122-year lifespan.

It's important to understand that scientific studies measure trends and reflect what happens to the norm. They don't tell what happens to everybody. Some people in their 70s and 80s function as well as those in their 30s and 40s. The wisdom and experience of older people often make up for deficits in performance.

The belief that pronounced and progressive mental decline is inevitable was and still is popular for several reasons. For one, until the 20th century, few people lived to healthy old ages. In 1900, when life expectancy was about 47 years, three million people, or four percent of the population, were older than age 65, and typically they were ill. In 1990, when life expectancy was more than 75 years, 30 million people, or 12 percent of the population, were older than age 65. A generation ago, frailty was seen among people in their 60s; today it is more typical among those in their 80s. Moreover, few people challenged the notion that aging meant inevitable brain decline because scientists knew little about the brain or the aging process. Today's understanding of how the normal brain ages comes from studies of the nervous system that began decades ago and are just now bearing results. Modern technologies now make it possible to explore the structure and functions of the brain in more depth than ever before and to ask questions about what actually happens in its aging cells.

Thus, neuroscientists are increasingly able to distinguish between the processes of normal aging and disease. While some changes do occur in normal aging, they are not as severe as scientists once thought.

All human behavior is determined by how well the brain's communication systems work. Often a failure in the cascade of one of these systems results in a disturbance of normal functions. Such a failure may be caused by an abnormal biochemical process or by a loss of neurons.

The cause of brain aging still remains a mystery. Dozens of theories abound. One says that specific "aging genes" are switched on at a certain time of life. Another points to genetic mutations or deletions. Other theories implicate hormonal influences, an immune system gone awry and the accumulation of damage caused by cell byproducts that destroy fats and proteins vital to normal cell function.

Aging neurons

The brain reaches its maximum weight near age 20 and slowly loses about 10 percent of its weight over a lifetime. Subtle changes in the chemistry and structure of the brain begin at midlife in most people. During a lifetime, the brain is at risk for losing some of its neurons, but neuron loss is not a normal process of aging. Brain tissue can respond to damage or loss of neurons in Alzheimer's disease or after stroke by expanding

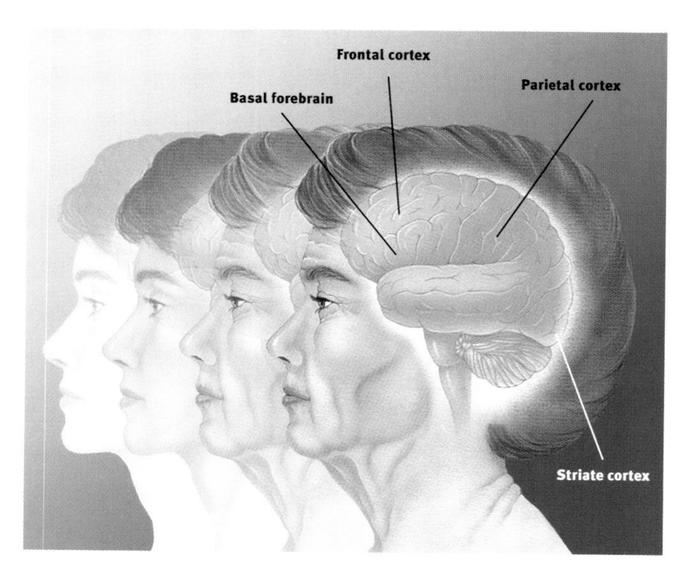

THE AGING BRAIN. Studies of people who have died contradict the popular belief that adults lose an enormous number of neurons every day. In fact, many areas of the brain, primarily in the cortex, maintain most of their neurons. Examples include the parietal cortex, which plays a role in sensory processes and language, and the striate cortex, which processes visual information. However, neurons in regions far below the cortex—such as the nucleus basalis, the principal component of the basal forebrain—decrease in number with age, and cell processes, such as axons and synapses, also change. The nucleus basalis sends connections to the cortex that produce acetylcholine, a chemical important for memory. Thus, these abnormalities may contribute to the mental declines that occur in some elderly individuals.

dendrites and refining connections between neurons. A damaged brain neuron can readjust to damage only if its cell body remains intact. If it does, regrowth can occur in dendrites and axons. When neurons are completely destroyed, nearby surviving neurons can compensate, in part, by growing new dendrites and connections.

Intellectual capacity

In the first large studies to follow the same group of normal healthy humans for many years, scientists have uncovered unexpected results. They report declines in some mental functions and improvements in others. In one study, the speed of carrying out certain tasks became slower, but vocabulary improved. Several studies found less severe declines in the type of intelligence relying on learned or stored information, compared with the type that uses the ability to deal with new information.

This research is supported by animal studies in which scientists found that changes in mental function are subtle. For example, in rodents and primates in which only minor brain abnormalities can be detected, certain spatial tasks, such as navigating to find food, tend to become more difficult with age.

The aging brain is only as resilient as its circuitry. Scientists debate whether this circuitry is changed only by neuron atrophy or whether some neuron loss over time also is inevitable. In any event, when the circuitry begins to break down, remaining neurons can respond by expanding their roles.

Learning conditions may dictate what happens to brain cells. Studies of rats shed light on some of the changes that occur in brain cells when the animals live in challenging and stimulating environments. In tests of middle-aged rats exposed to these environments, researchers found that dendrites in the cerebral cortex, which is responsible for all conscious activity, developed more and longer branches when compared with rats housed in isolated conditions. Another study showed that brain cells in rats given acrobatic training had greater numbers of synapses per cell than rats given only physical exercise or rats who were inactive. The scientists concluded that motor learning generates new synapses. Physical exercise, however, improved blood circulation in the brain.

Other scientists report that rats reared in a stimulating environment made significantly fewer errors on a maze test than did similar rats kept in an isolated environment. Moreover, the stimulated rats showed an increase in brain weight and cortical thickness when compared with the control animals.

Older rats tend to form new dendrites and synapses as do younger animals in response to enriched environments. But the response is more sluggish and not as large. Compared to younger rats, the older rats have less growth of the new blood vessels that nourish neurons.

While much has been learned about the aging brain, many questions remain to be answered. For instance, does the production of proteins decline with age in all brain neurons? In a given neuron, does atrophy cause a higher likelihood of death? How does aging affect gene expression in the brain—the organ with the greatest number of active genes? Are there gender differences in brain aging that may be due to hormonal changes at menopause?

Neuroscientists speculate that certain genes may be linked to events leading to death in the nervous system. By understanding the biology of the proteins produced by genes, scientists hope to be able to influence the survival and degeneration of neurons.

Advances

Parkinson's disease. This neurologic disorder afflicts one million individuals in the U.S., the majority of whom are older than 50. Parkinson's is characterized by symptoms of slowness of movement, muscular rigidity and tremor.

The discoveries in the late 1950s that the level of dopamine was decreased in the brains of patients was followed in the 1960s by the successful treatment of this disorder by administration of the drug levodopa. This drug is changed to dopamine in the brain. The successful treatment of Parkinson's by replacement therapy is one of the greatest success stories in all of neurology. Levodopa is now combined with another drug, carbidopa, that reduces the peripheral breakdown of levodopa, thus allowing greater levels to reach the brain and reducing side effects. Also playing an important role are newer drugs such as pergolide that act directly upon dopamine receptors and other inhibitors of dopamine breakdown.

Genetic studies have demonstrated several inheritable gene abnormalities in certain families, but the vast majority of cases of Parkinson's occur sporadically. It is believed that heredity factors may render some individuals more vulnerable to environmental factors such as pesticides. The discovery in the late 1970s that a chemical substance, MPTP, can cause parkinsonism in drug addicts stimulated intensive research on the causes of the disorder. MPTP was accidently synthesized by illicit drug designers seeking to produce a heroin-like compound. MPTP was found to be converted in the brain to a substance that destroys dopamine neurons. Parkinson's is now being intensively studied in a primate MPTP model.

In the past several decades, scientists have shown in a primate model of Parkinson's that specific regions in the basal ganglia, the collections of cell bodies deep in the brain, are abnormally overactive. Most importantly, they found that surgical destruction of these overactive nuclei—the pallidum and subthalamic nucleus—can greatly reduce symptoms. The past decade has witnessed a resurgence in this surgical procedure, *pallidotomy,* and more recently *chronic deep brain stimulation.* These techniques are highly successful for treating patients who have experienced significant worsening of symptoms and are troubled by the development of drug-related involuntary movements. The past decade has also seen further attempts to treat such patients with surgical implantation of cells, such as fetal cells, capable of producing dopamine. Replacement therapy with stem cells also is being explored.

Pain

If there is a universal experience, pain is it. Each year, more than 97 million Americans suffer chronic, debilitating headaches or a bout with a bad back or the pain of arthritis—all at a total cost of some $100 billion. But it need not be that way. New discoveries about how chemicals in the body transmit and intercept pain have paved the way for new treatments for both chronic and acute pain.

Until the middle of the 19th century, pain relief during surgery relied on natural substances, such as opium, alcohol and cannabis. All were inadequate and short-lived. Not until 1846 did doctors discover the anesthetic properties of ether, first in animals and then in humans. Soon afterwards, the usefulness of chloroform and nitrous oxide became known and heralded a new era in surgery. The dozens of drugs used today during surgery abolish pain, relax muscles and induce unconsciousness. Other agents reverse these effects.

Local anesthesia is used in a limited area of a person's body to prevent pain during examinations, diagnostic procedures, treatments and surgical operations. The most famous of these agents, which temporarily interrupt the action of pain-carrying nerve fibers, is *Novocain.* Until recently, Novocain was used as a local anesthetic by dentists; *lidocaine* is more popular today.

Analgesia produces loss of pain sensation without loss of sensitivity to touch. The two main types of analgesics are nonopioids (aspirin and related non-steroidal anti-inflammatory drugs such as ibuprofen, naproxen and acetaminophen) and opioids (morphine, codeine). Nonopioid analgesics are useful for treating mild or moderate pain, such as headache or toothache. Moderate pain also can be treated by combining a mild opioid, such as codeine with aspirin. Opioids are the most potent painkillers and are used for severe pain, such as that occurring after major chest or abdominal surgery.

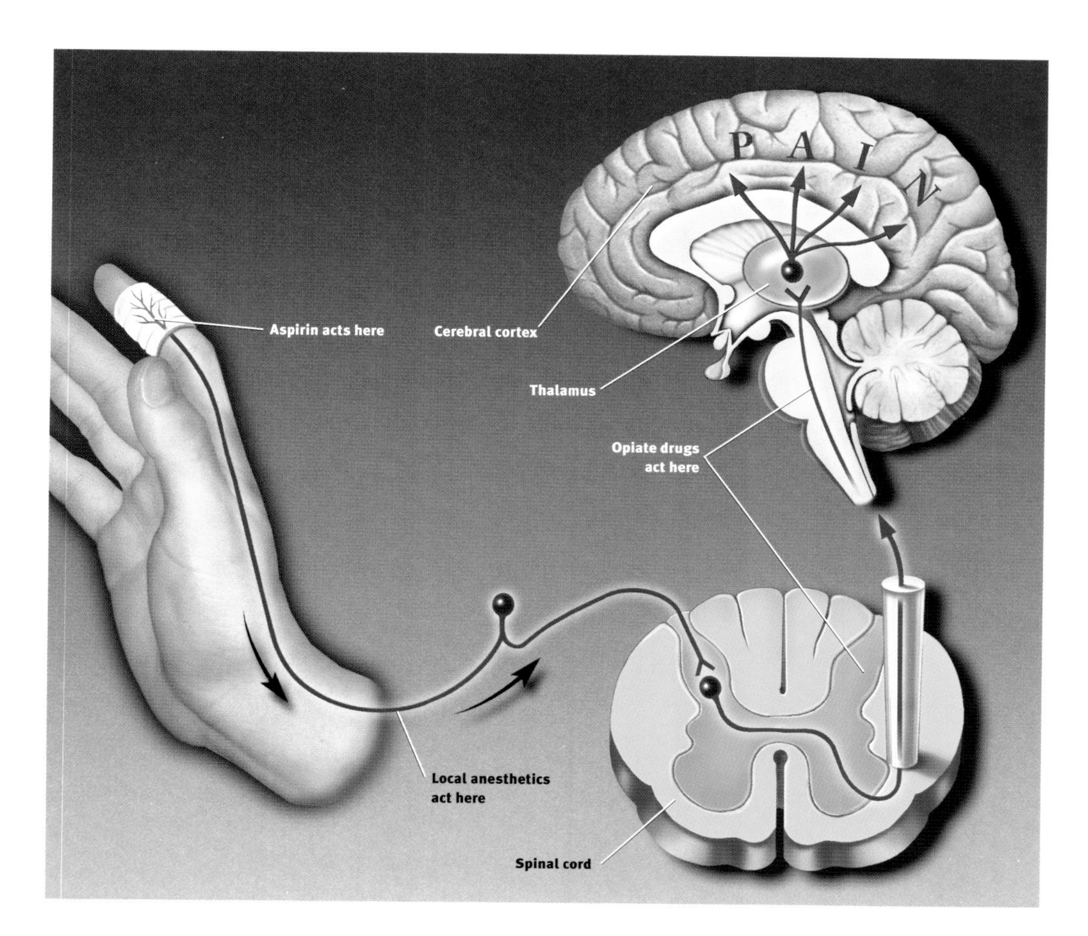

HOW PAIN KILLERS WORK. At the site of injury, the body produces prostaglandins that increase pain sensitivity. Aspirin, which acts primarily in the periphery, prevents the production of prostaglandins. Acetaminophen is believed to block pain impulses in the brain itself. Local anesthetics intercept pain signals traveling up the nerve. Opiate drugs, which act primarily in the central nervous system, block the transfer of pain signals from the spinal cord to the brain.

Insights into the body's own pain-control system mediated by naturally occurring opioids led to the use of injections of morphine and endorphins, and other opioids, into the cerebrospinal fluid in which the spinal cord is bathed without causing paralysis, numbness or other severe side effects. This technique came about through experiments with animals that first showed that injecting opioids into the spinal cord could produce profound pain control. This technique is now commonly used in humans to treat pain after surgery.

New knowledge about other receptors and chemical mediators involved in the transmission of pain are leading to the development of new approaches to managing pain. These include drugs that intercept pain messages at receptors that bind glutamate, the major excitatory neurotransmitter in pain pathways. Other studies are using molecular biology techniques to identify specialized receptors and ion channels in nerve endings that signal tissue damage of the skin, muscle or viscera. These studies have the promise of leading to new classes of analgesic agents in the future.

Epilepsy

A chronic neurological disorder characterized by sudden, disorderly discharge of brain cells, epilepsy is marked by recurrent seizures that temporarily alter one or more brain functions. The disorder affects approximately one percent of the population.

Many different forms of epilepsy have been recognized. Epilepsy, which can start at any age, can result from inheriting a mutant gene. It also can result from a wide variety of diseases or injuries (including head injury), birth trauma, brain infection (such as meningitis), brain tumors, stroke, drug intoxication, drug or alcohol withdrawal states and metabolic disorders. More than a dozen mutant genes that cause human epilepsy have been identified during the past decade. In 70 percent of cases, however, the cause is unknown.

Seizures are of two types. *Generalized seizures*, which result in loss of consciousness, can cause several behavioral changes including convulsions or sudden changes in muscle tone and arise when there is excessive electrical activity over a wide area of the brain. *Partial seizures* may occur in full consciousness or with altered awareness, and can cause behavioral changes. They can range from visual, auditory and sensory disturbances to uncontrolled movements, and arise from excessive electrical activity in a limited area of the brain.

The drug *phenytoin* was a major advance in the treatment of epilepsy because it illustrated that antiseizure medications need not cause sedation (as does *phenobarbital*, an older drug for epilepsy) and encouraged the search for other drugs. Today more than a dozen medications, approximately half of which were introduced in the last several years, are used to prevent seizures. The principal targets of antiseizure drugs are voltage-gated ion channels permeable to sodium or calcium and synapses using the transmitter GABA, a naturally occurring substance in the brain that acts to inhibit electrical discharge. Identification of the mutant genes underlying human epilepsy is providing new targets for the next generation of antiseizure drugs.

In many instances, epilepsy can be controlled with a single antiseizure drug that lessens the frequency of seizures. Sometimes a combination of drugs is necessary. Complete control of seizures can be achieved in more than 50 percent of patients, and another 25 percent can be improved significantly. It is hoped that the newly available antiseizure drugs will provide complete control in additional patients.

Surgery, considered for the patients who do not respond to drugs, should be performed only at specialized medical centers qualified to handle epilepsy surgery. One type of surgery requires precise location and removal of the area of the brain where the seizures originate. About 90 percent of properly selected patients experience striking improvement or complete remission of seizures. Another type of surgery separates the left and right hemispheres of the brain to control a type of seizure that originates in one hemisphere and spreads to involve the whole brain.

A new form of epilepsy treatment, electrical stimulation therapy, was introduced during the mid-1990s as another option for hard-to-control seizures. The implantable pacemaker-like device delivers small bursts of electrical energy to the brain via the vagus nerve on the side of the neck.

Major depression

This affliction, with its harrowing feelings of sadness, hopelessness, pessimism, loss of interest in life and reduced emotional wellbeing, is one of the most common and debilitating mental disorders. Depression is as disabling as coronary disease or arthritis. Depressed individuals are 18 times more likely to attempt suicide than people with no mental illness.

Annually, major depression affects five percent of the population or 9.8 million Americans aged 18 years and older. Fortunately, 80 percent of patients respond to drugs, psychotherapy or a combination of the two. Some severely depressed patients can be helped with electroconvulsive therapy.

Depression arises from many causes: biological (including genetic), psychological, environmental or a combination of these. Stroke, hormonal disorders, antihypertensives and birth control pills also can play a part.

Physical symptoms—disturbances of sleep, sex drive, appetite and digestion—are common. Some of these symptoms may reflect the fact that the disorder affects the delicate hormonal feedback system linking the hypothalamus, the pituitary gland and the adrenal glands. For example, many depressed patients secrete excess cortisol, a stress hormone, and do not respond appropriately to a hormone that should counter cortisol suppression. When tested in sleep laboratories, depressed patients' electroencephalograms (EEGs) often exhibit abnormalities in their sleep patterns.

The modern era of drug treatment for depression began in the late 1950s. Most antidepressants affect norepinephrine and serotonin in the brain, apparently by correcting the abnormal excess or inhibition of the signals that control mood, thoughts, pain and other sensations. The tricyclic antidepressants primarily block the reabsorption and inactivation of serotonin and norepinephrine to varying degrees.

Another class of antidepressant medications is the *monoamine oxidase inhibitors* (MAOIs). MAOIs are thought to be more complicated than tricyclics. These agents inhibit *monamine oxidase*, an enzyme that breaks down serotonin, norepinephrine and dopamine, allowing these chemicals to remain active. During the 1950s, the first of the MAOIs, iproniazid, was found to make experimental animals hyperalert and hyperactive. By 1957, scientists had proven iproniazid's benefit in patients. Later, other MAOIs were developed. Today, three are available for use: *isocarboxazid, phenelzine and tranylcypromine*.

The popular medication *fluoxetine* (Prozac) is the first of a new class of drugs, *serotonin reuptake inhibitors*. Fluoxetine blocks the reabsorption and inactivation of serotonin and keeps it active in certain brain circuits. This seems to restore overall serotonin activity to a more normal state and ease depression.

Manic-depressive illness

Patients with manic-depressive illness, also known as bipolar disorder, usually experience episodes of deep depression and manic highs, with a return to relatively normal functioning in between. They also have an increased risk of suicide. Manic depression affects 1.2 percent of Americans age 18 or older annually, or 2.2 million individuals. Approximately equal numbers of men and women suffer from this disorder.

Manic-depressive disorder tends to be chronic, and episodes can become more frequent without treatment. Because manic depression runs in families, efforts are under way to identify the responsible gene or genes.

However, manic-depressive patients also can benefit from a broad array of treatments. One of these is lithium. During the 1940s, researchers showed that lithium injections into guinea pigs made them placid, which implied mood-stabilizing effects. When given to manic patients, lithium calmed them and enabled them to return to work and live relatively normal lives. Regarded as both safe and effective, lithium is often used to prevent recurrent episodes.

Other useful medications include anticonvulsants, such as valproate or carbamazepine, which can have mood-stabilizing effects and may be especially useful for difficult-to-treat bipolar episodes. Newer anticonvulsant medications, are being studied to determine how well they work in stabilizing mood cycles.

Challenges

Addiction. Drug abuse is one of the nation's most serious health problems. Indeed, six percent of Americans, roughly 15 million people, abuse drugs on a regular basis. Recent estimates show that the abuse of drugs, including alcohol and nicotine from tobacco, costs the nation more than $276 billion each year.

If continued long enough, *drug abuse*—often defined as harmful drug use—can eventually alter the very structure of the brain, producing a true brain disorder. This disorder is called *drug addiction* or *drug dependence*. Drug addiction is defined as having lost much control over drug taking, even in the face of adverse physical, personal or social consequences.

People abuse drugs for a simple reason: Drugs produce feelings of pleasure, or they remove feelings of stress and emotional pain. Neuroscientists have found that almost all abused drugs produce pleasure by activating a specific network of neurons called the brain reward system. The circuit is normally involved in an important type of learning that helps us to stay alive. It is activated when we fulfill survival functions, such as eating when we are hungry or drinking when we are thirsty. In turn, our brain rewards us with pleasurable feelings that teach us to repeat the task. Because drugs inappropriately turn on this reward circuit, people want to repeat drug use.

Neuroscientists have also learned specifically how drugs affect neurons to exert their actions. Neurons release special chemicals, called neurotransmitters, to communicate with each other. Drugs of abuse alter the ways in which neurotransmitters carry their messages from neuron to neuron. Some drugs mimic neurotransmitters while others block them. Still others alter the way that the neurotransmitters are released or inactivated. The brain reward system is inappropriately activated because drugs alter the chemical messages sent among neurons in this circuit.

Finally, neuroscientists also have learned that addiction requires more than the activation of the brain reward system. The process of becoming addicted appears to be influenced by many factors. Motivation for drug use is an important one. For example, people who take drugs to get high may get addicted, but people who use them properly as medicine rarely do. Also genetic susceptibility or environmental factors, like stress, may alter the way that people respond to drugs. In addition, the development of *tolerance*—the progressive need that accompanies chronic use for a higher drug dose to achieve the same effect—varies in different people. So does *drug dependence*—the adaptive physiological state that results in withdrawal symptoms when drug use stops. While tolerance and dependence are standard responses of the brain and body to the presence of drugs, addiction requires that these occur while a *motivational form of dependence*—the feeling that a person can't live without a drug, accompanied by negative affective states—is also developing. Together these insights on abuse and addiction are leading to new therapies.

Nicotine Some 57 million Americans were current smokers in 1999, and another 7.6 million used smokeless tobacco, making nicotine one of the most widely abused substances. Tobacco kills more than 430,000 U.S. citizens each year—more than alcohol, cocaine, heroin, homicide, suicide, car accidents, fire, and AIDS combined. Tobacco use is the leading preventable cause of death in the United States. Smoking is responsible for approximately seven percent of total U.S. health care costs, an estimated $80 billion each year. The direct and indirect costs of smoking are estimated at more than $138 billion per year.

Nicotine acts through the well known cholinergic nicotinic receptor. This drug can act as both a stimulant and a sedative. Immediately after exposure to nicotine, there is a "kick" caused in part by the drug's stimulation of the adrenal glands and resulting discharge of epinephrine. The rush of adrenaline stimulates the body and causes a sudden release of glucose as well as an increase in blood pressure, respiration and heart rate. Nicotine also suppresses insulin output from the pancreas, which means that smokers are always slightly hyperglycemic. In addition, nicotine indirectly causes a release of dopamine in the brain regions that control pleasure and motivation. This is thought to underlie the pleasurable sensations experienced by many smokers.

Much better understanding of addiction, coupled with the identification of nicotine as an addictive drug, has been instrumental in developing treatments. Nicotine gum, the transder-

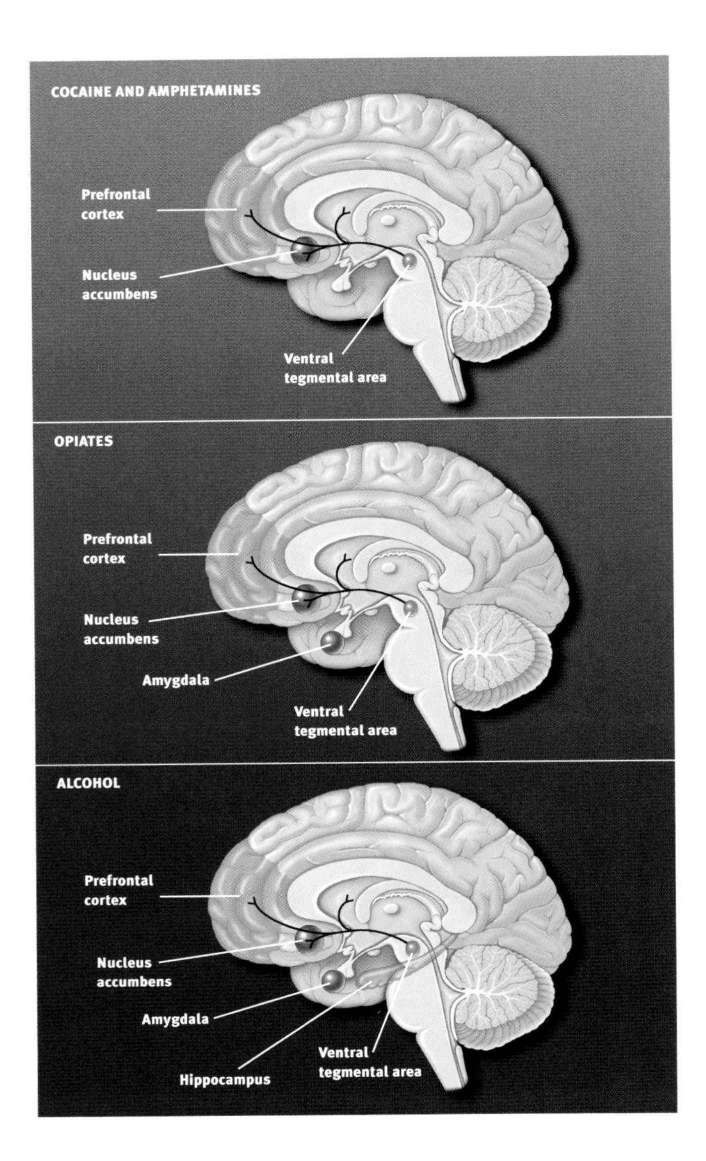

BRAIN DRUG REWARD SYSTEMS. Scientists are not certain about all the structures involved in the human brain reward system. However, studies of rat and monkey brains, and brain imaging studies in humans, have provided many clues. These illustrations show what areas are most likely part of the reward systems in the human brain. A central group of structures is common to the actions of all drugs. These structures include a collection of dopamine-containing neurons found in the ventral tegmental area. These neurons are connected to the nucleus accumbens and other areas, such as the prefrontal cortex. Cocaine exerts its effects mainly through this system. Opiates act in this system and many other brain regions, including the amygdala, that normally use opioid peptides. Opioids are naturally occurring brain chemicals that induce the same actions as drugs, such as heroin and morphine. Alcohol activates the core reward system and additional structures throughout the brain because it acts where GABA and glutamate are used as neurotransmitters. GABA and glutamate are widely distributed in the brain, including the cortex, hippocampus, amygdala and nucleus accumbens.

mal patch, nasal spray and inhaler all appear to be equally effective in successfully treating more than one million people addicted to nicotine. These techniques are used to relieve withdrawal symptoms, produce less severe physiological alterations than tobacco-based systems and generally provide users with lower overall nicotine levels than they receive with tobacco. The first non-nicotine prescription drug, bupropion, an antidepressant marketed as Zyban, has been approved for use as a pharmacological treatment for nicotine addiction. Behavioral treatments are important for helping an individual learn coping skills for both short- and long-term prevention of relapse.

Psychostimulants In 1997, 1.5 million Americans were current cocaine users. A popular, chemically altered form of cocaine, crack, is smoked. It enters the brain in seconds, producing a rush of euphoria and feelings of power and self-confidence. The key biochemical factor that underlies the reinforcing effects of psychostimulants is the brain chemical dopamine. We feel pleasure-like effects when dopamine-containing neurons release dopamine into specific brain areas that include a special portion of the nucleus accumbens. Cocaine and amphetamines produce their intense feelings of euphoria by increasing the amount of dopamine that is available to send messages within the brain reward system.

Cocaine users often go on binges, consuming a large amount of the drug in just a few days. A "crash" occurs after this period of intense drug-taking and includes symptoms of emotional and physical exhaustion and depression. These symptoms may result from an actual crash in dopamine function and the activity of another brain chemical, serotonin, as well as an increase in the response of the brain systems that react to stress. Vaccines to produce antibodies to cocaine in the bloodstream are in clinical trials.

Opiates Humans have used opiate drugs, such as morphine, for thousands of years. Monkeys and rats readily self-administer heroin or morphine and, like humans, will become tolerant and physically dependent with unlimited access. Withdrawal symptoms range from mild flu-like discomfort to major physical ailments, including severe muscle pain, stomach cramps, diarrhea and unpleasant mood.

Opiates, like psychostimulants, increase the amount of dopamine released in the brain reward system and mimic the effects of endogenous opioids such as opioid peptides. Heroin injected into a vein reaches the brain in 15 to 20 seconds and binds to opiate receptors found in many brain regions, including the reward system. Activation of the receptors in the reward circuits causes a brief rush of intense euphoria, followed by a couple of hours of a relaxed, contented state.

Opiates create effects like those elicited by the naturally occurring opioid peptides. They relieve pain, depress breathing, cause nausea and vomiting, and stop diarrhea—important medical uses. In large doses, heroin can make breathing shal-

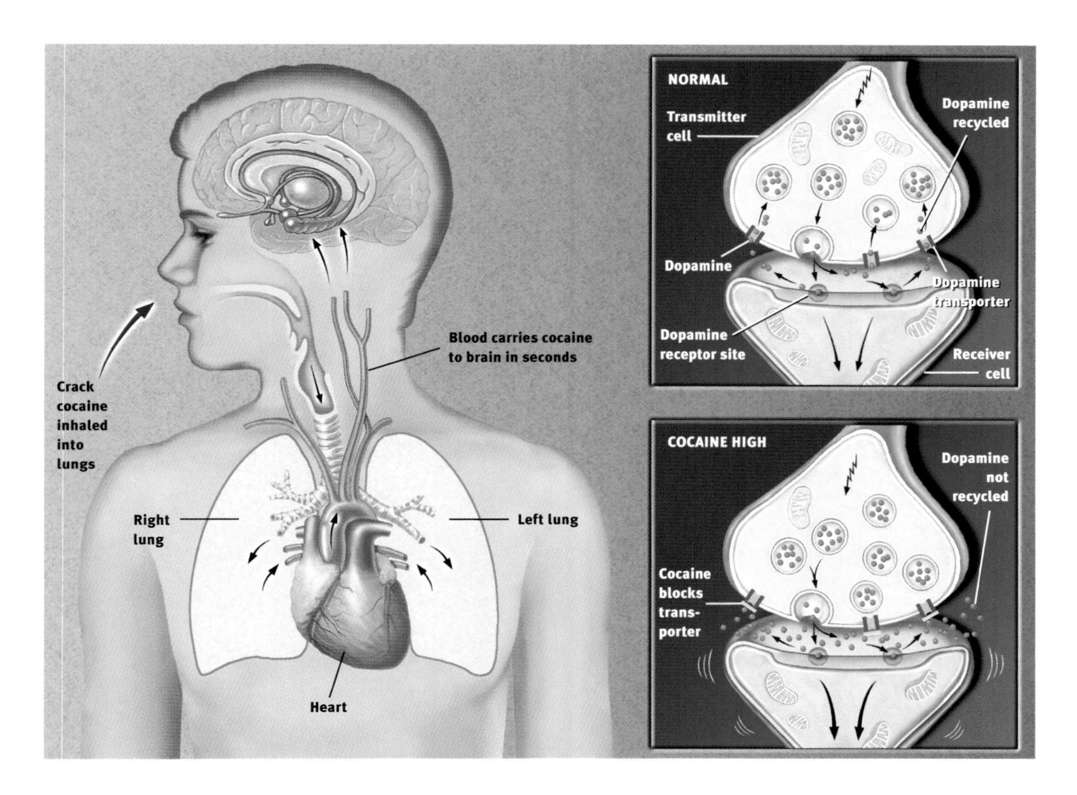

HOW CRACK COCAINE AFFECTS THE BRAIN. Crack cocaine takes the same route as nicotine by entering the bloodstream through the lungs. Within seconds, it is carried by the blood to the brain. The basis for increased pleasure occurs at the gap where the impulses that represent neural messages are passed from one neuron to another. This gap is called a synapse. Dopamine-containing neurons normally relay their signals by releasing dopamine into many synapses. Dopamine crosses the synapse and fits into receptors on the surface of the receiving cell. This triggers an electrical signal that is relayed through the receiver. Then, to end the signal, dopamine molecules break away from the receptors and are pumped back into the nerve terminals that released them. Cocaine molecules block the pump or "transporter," causing more dopamine to accumulate in the synapse. Pleasure circuits are stimulated again and again, producing euphoria.

low or stop altogether—the cause of death in thousands of people who have died of a heroin overdose.

A standard treatment for opiate addiction involves methadone, a long-acting oral opiate that helps keep craving, withdrawal and relapse under control. Methadone helps opiate addicts rehabilitate themselves by preventing withdrawal symptoms that are powerful motivators of drug use. A synthetic opiate, known as LAAM, can block the effects of heroin for up to 72 hours with minimal side effects when taken orally. In 1993 the Food and Drug Administration approved the use of LAAM for treating patients addicted to heroin. Its long duration of action permits dosing just three times per week, thereby eliminating the need for daily dosing. LAAM will be increasingly available in clinics that already dispense methadone. Naloxone and naltrexone are medications that also block the effects of morphine, heroin and other opiates. As *antagonists*, they are especially useful as antidotes. Another medication to treat heroin addiction, buprenorphine, causes weaker opiate effects and is less likely to cause overdose problems.

Alcohol Although legal, alcohol is highly addictive. Alcohol abuse and alcohol addiction (sometimes referred to as alcoholism or alcohol dependence) are the nation's major drug problem, with some people being more susceptible to them than others. Nearly 14 million abuse alcohol or are alcoholic. *Fetal alcohol syndrome* is the leading preventable cause of mental retardation. It affects about 0.5 to 3 of every 1,000 babies born in the United States. Chronic liver diseases, including cirrhosis—the main chronic health problem associated with alcohol addiction—are responsible for more than 25,000 deaths each year. The annual cost of alcohol abuse and addiction is estimated at $185 billion.

Genetic and environmental factors contribute to alcoholism, but no single factor or combination of factors enables doctors to predict who will become an alcoholic.

Ethanol, the active ingredient in alcoholic beverages, reduces anxiety, tension and inhibitions. In low doses it may act as a stimulant, whereas at higher doses, it acts as a depressant. In both cases, it significantly alters mood and behavior. It can also cause heat loss and dehydration.

The drug, which is easily absorbed into the bloodstream and the brain, affects several neurotransmitter systems. For example, alcohol's interaction with the GABA receptor can calm anxiety, impair muscle control and delay reaction time. At higher doses, alcohol also decreases the function of NMDA receptors that recognize the neurotransmitter glutamate. This interaction can cloud thinking and eventually lead to a coma.

Researchers are developing treatments, which interfere with molecules, such as the opioid peptides, and trigger alcohol's positive reinforcing effects. One such drug, naltrexone, recently has been approved for treating alcoholism.

Marijuana This drug can distort perception, and alter the sense of time, space and self. In certain situations, marijuana can produce intense anxiety.

In radioactive tracing studies, scientists found that *tetrahydrocannabinol* (THC), the active ingredient in marijuana, binds to specific receptors, many of which coordinate movement. This may explain why people who drive after they smoke marijuana are impaired. The hippocampus, a structure involved with memory storage and learning, also contains many THC receptors. This may explain why heavy users or those intoxicated on marijuana have poor short-term memory and problems processing complex information. Scientists recently discovered that these receptors normally bind to a natural internal chemical called *anandamide*, and are now working to see how this natural marijuana affects brain function.

Club Drugs Ecstasy, Herbal Ecstasy, Rohypnol, GHB and ketamine are among the drugs used by some teens and young adults as part of rave and trance events which are generally night-long dances, often held in warehouses. The drugs are rumored to increase stamina and to produce intoxicating highs that are said to deepen the rave or trance experience. Recent hard science, however, is showing serious damage to several parts of the brain from use of some of these drugs.

Many users tend to experiment with a variety of club drugs in combination. This creates a larger problem because combinations of any of these drugs, particularly with alcohol, can lead to unexpected adverse reactions and even death after high doses. Physical exhaustion can enhance some toxicities and problems.

MDMA, called "Adam," "Ecstasy," or "XTC," on the street, is a synthetic, psychoactive drug with hallucinogenic and amphetamine-like properties. Users encounter problems similar to those found with the use of amphetamines and cocaine. Recent research also links Ecstasy use to long-term damage to those parts of the brain critical to thought, memory and pleasure.

Rohypnol, GHB, and ketamine are predominantly central nervous system depressants. Because they are often colorless, tasteless and odorless, they can be easily added to beverages and ingested unknowingly. These drugs have emerged as the so-called "date rape" drugs. When mixed with alcohol, Rohypnol can incapacitate a victim and prevent them from resisting sexual assault. Also, Rohypnol may be lethal when mixed with alcohol and other depressants. Since about 1990, GHB (gamma hydroxy-butyrate) has been abused in the U.S. for euphoric, sedative and anabolic (body building) effects. It, too, has been associated with sexual assault. Ketamine is another central nervous system depressant abused as a "date rape" drug. Ketamine, or "Special K," is a rapid-acting general anesthetic. It has sedative-hypnotic, analgesic and hallucinogenic properties. It is marketed in the U.S. and a number of foreign countries for use as a general anesthetic in both human and veterinary medical practice.

Alzheimer's disease

One of the most frightening and devastating of all neurological disorders, is the dementia that occurs in the elderly. The most common cause of this illness is Alzheimer's disease (AD). Rare before age 60 but increasingly prevalent in each decade thereafter, AD affects an estimated 4 to 5 million Americans. By the year, 2040, it is predicted to affect approximately 14 million individuals in the U.S.

The earliest symptoms are forgetfulness and memory loss; disorientation to time or place; and difficulty with concentration, calculation, language and judgment. Some patients have severe behavioral disturbances and may even become psychotic. The illness is progressive. In the final stages, the affected individual is incapable of self-care. Unfortunately, no effective treatments exist, and patients usually die from pneumonia or some other complication. AD, which kills 100,000 people a year, is one of the leading causes of death in the U.S.

In the earliest stages, the clinical diagnosis of possible or probable AD can be made with greater than 80 percent accuracy. As the course of the disease progresses, the accuracy of diagnosis at Alzheimer's research centers exceeds 90 percent. The diagnosis depends on medical history, physical and neurological examinations, psychological testing, laboratory tests and brain imaging studies. At present, final confirmation of the diagnosis requires examination of brain tissue, usually obtained at autopsy.

The causes and mechanisms of the brain abnormalities are not yet fully understood, but great progress has been made through genetics, biochemistry, cell biology and experimental treatments. Microscopic examination of AD brain tissue shows abnormal accumulations of a small fibrillar peptide, termed a *beta amyloid*, in the spaces around synapses (*neuritic plaques*), and by abnormal accumulations of a modified form of the protein tau in the cell bodies of neurons (*neurofibrillary tangles*). The plaques and tangles are mostly in brain regions important for memory and intellectual functions.

In cases of AD, there are reductions in levels of markers for several neurotransmitters, including acetylcholine, somatostatin, monoamine and glutamate, that allow cells to communicate with one another. Damage to these neural systems, which are critical for attention, memory, learning and higher cognitive abilities, is believed to cause the clinical symptoms.

Approximately five percent to 10 percent of individuals with

AD have an inherited form of the disease. These patients often have early-onset illness. Recently, scientists have identified mutations in AD-linked genes on three chromosomes. The gene encoding the *amyloid precursor protein* gene is on chromosome 21. In other families with early-onset AD, mutations have been identified in the presenilin 1 and 2 genes, which are on chromosomes 14 and 1, respectively. *Apolipoprotein* E (apoE), a chromosome 19 gene, which influences susceptibility in late life, exists in three forms, with apoE4 clearly associated with enhanced risk.

Treatments are available mostly only for some symptoms of AD, such as agitation, anxiety, unpredictable behavior, sleep disturbances and depression. Three drugs treat cognitive symptoms in patients with mild to moderate Alzheimer's. These agents improve memory deficits temporarily and modestly in 20 percent to 30 percent of patients. Several other approaches, such as antioxidants, anti-inflammatories and estrogens, are being tested.

An exciting new area of research is the use of approaches in which genes are introduced in mice. These transgenic mice carrying mutant genes linked to inherited AD develop behavioral abnormalities and some of the cellular changes that occur in humans. It is anticipated that these mice models will prove very useful for studying the mechanisms of AD and testing novel therapies.

Moreover, researchers have begun to knock out genes playing critical roles in the production of amyloid. These enzymes, termed beta and gamma secretase, which cleave the amyloid peptide from the precursor, are clearly targets for development of drugs to block amyloid.

Learning disorders

An estimated 10 percent of the population, as many as 25 million Americans, have some form of learning disability involving difficulties in the acquisition and use of listening, speaking, reading, writing, reasoning or mathematical abilities. They often occur in people with normal or high intelligence.

Dyslexia, or specific reading disability, is the most common and most carefully studied of the learning disabilities. It affects 80 percent of all of those identified as learning-disabled. Dyslexia is characterized by an unexpected difficulty in reading in children and adults who otherwise possess the intelligence, motivation and schooling considered necessary for accurate and fluent reading.

Previously, it was believed that dyslexia affected boys primarily. However, more recent data indicate similar numbers of boys and girls are affected. Studies indicate that dyslexia is a persistent, chronic condition. It does not represent a transient "developmental lag."

There is now a strong consensus that the central difficulty in dyslexia reflects a deficit within the language system, and more specifically, in a component of the language system called phonology. This is illustrated in difficulty transforming the letters on the page to the sound structure of the language.

As children approach adolescence, a manifestation of dyslexia may be a very slow reading rate. Children may learn to read words accurately but they will not be fluent or automatic, reflecting the lingering effects of a phonologic deficit. Because they are able to read words accurately (albeit very slowly), dyslexic adolescents and young adults may mistakenly be assumed to have "outgrown" their dyslexia. The ability to read aloud accurately, rapidly and with good expression as well as facility with spelling may be most useful clinically in distinguishing students who are average from those who are poor readers.

A range of investigations indicates that there are differences in the temporo-parieto-occipital brain regions between dyslexic and non-impaired readers. Recent data using functional brain imaging indicate that dyslexic readers demonstrate a functional disruption in an extensive system in the posterior portion of the brain. The disruption occurs within the neural systems linking visual representations of the letters to the phonologic structures they represent. The specific cause of the disruption in neuronal systems in dyslexia is not entirely understood. However, it is clear that dyslexia runs in families and can be inherited.

Interventions to help children with dyslexia focus on teaching the child that words can be segmented into smaller units of sound and that these sounds are linked with specific letters and letter patterns. In addition, children with dyslexia require practice in reading stories, both to allow them to apply their newly acquired decoding skills to reading words in context and to experience reading for meaning.

Stroke

Until recently, if you or a loved one had a stroke, your doctor would tell your family there was no treatment. In all likelihood, the patient would live out the remaining months or years with severe neurological impairment.

This dismal scenario has now been radically altered. For one, use of the clot-dissolving bioengineered drug, tissue plasminogen activator (tPA), is now a standard treatment in many hospitals. This approach rapidly opens blocked vessels to restore circulation before oxygen loss causes permanent damage. Given within three hours of a stroke, it often can help in returning patients to normal.

Also, attitudes about the nation's third leading cause of death are changing rapidly. Much of this has come from new and better understandings of the mechanisms that lead to the death of neurons following stroke and devising ways to protect these neurons. A variety of chemicals appear to play a role, including calcium, potassium and zinc, and may be important in devising new treatments.

Stroke affects roughly 700,000 Americans a year — 150,000 of whom die; total annual costs are estimated at $43 billion.

STROKE. A stroke occurs when a blood vessel bringing oxygen and nutrients to the brain bursts or is clogged by a blood clot. This lack of blood can cause cell death within minutes. One theory is that the overexcited dying nerve cells release neurotransmitters, especially glutamate, onto nearby nerve cells. These nearby nerve cells become overexcited and overloaded with calcium and die. This is one of the places where scientists think they may be able to intervene to stop the process of cell death. Depending on its location, a stroke can have different symptoms. They include paralysis on one side of the body or a loss of speech. The effects of stroke are often permanent because dead brain cells are not replaced.

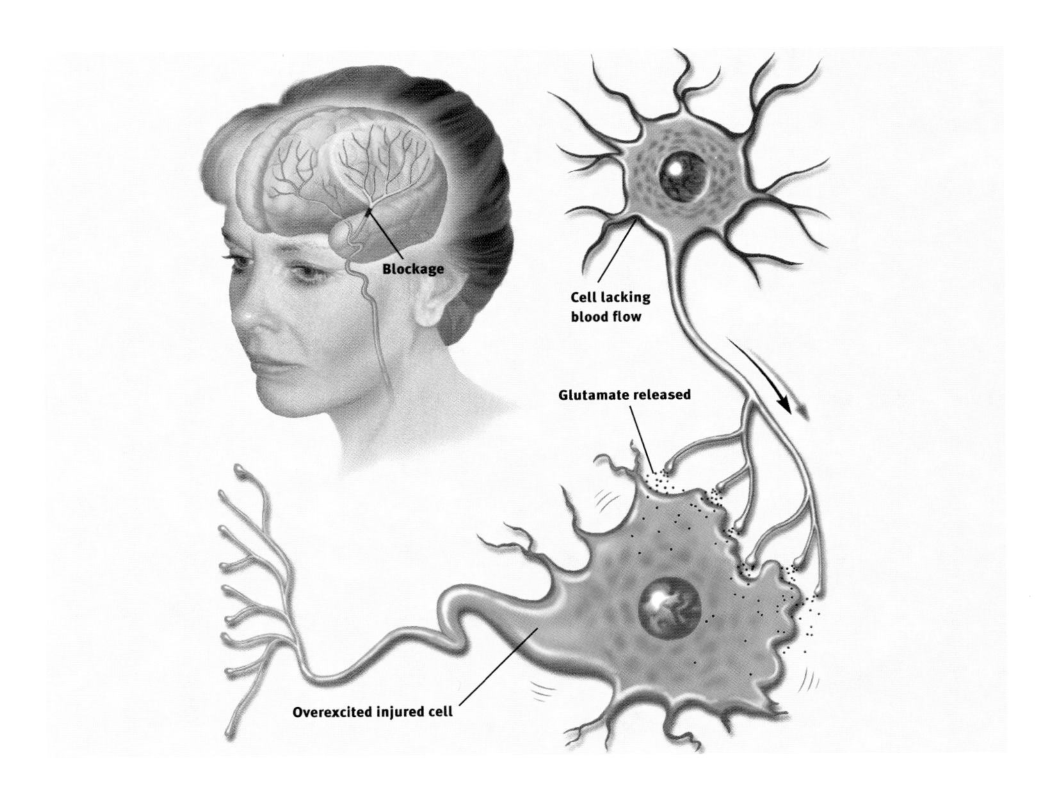

A stroke occurs when a blood vessel bringing oxygen and nutrients to the brain bursts or is clogged by a blood clot or some other particle. This deprives the brain of blood, causing the death of neurons within minutes. Depending on its location, a stroke can cause many permanent disorders, such as paralysis on one side of the body and loss of speech.

Stroke often occurs in individuals over 65 years of age, yet a third are younger. Stroke tends to occur more in males and blacks and in those with diabetes, high blood pressure, heart disease, obesity, high cholesterol and a family history of stroke.

In addition to tPA, increased use of preventive measures are battling the disorder. Controlling risk factors such as obesity, blood pressure, diabetes and high cholesterol can help prevent stroke. Other specific treatments involving surgery can clear clogs in the arteries of the neck region and help prevent a cutoff of blood supply.

Treatments that target the heart's blood flow can prevent stroke. Surgery can help repair damaged heart valves. Drugs can reduce the chance of clots forming, traveling to the brain and causing a stroke.

Other experimental therapies under investigation may lead to even bigger payoffs for patients in the future. Some strategies target mechanisms inside the neuron. In this way, the vicious cycle of local damage followed by a widening fringe of biochemical-induced neuronal death can be slowed. A number of classes of drugs have been shown to be effective in animal studies.

Another promising possibility is the use of neural stem cells. Some animal studies have shown that an injection of stem cells aids recovery even if administered as long as a day after the injury. Stem cells administered along with a growth factor resulted in greater improvement than with either treatment alone. The double regimen resulted in behavioral improvement as well as decreased stroke-induced brain loss.

Neurological trauma

A magic bullet has not been found, but doctors have discovered several methods to stave off severe neurological damage caused by head and spinal cord injuries. These treatments include better emergency care, a drug to help reduce damage and improved rehabilitation techniques.

Some 500,000 people suffer traumatic head injuries requiring hospitalization each year, and roughly 100,000 die—many before reaching the hospital. Economic costs approach $25 billion annually.

Greater use of *computed tomography* (CT) and *magnetic resonance imaging* (MRI) allows doctors more readily to see potentially life-threatening swelling and act immediately more readily. Doctors can bore a small hole in the skull and insert a tube attached to a pressure monitor. When the intracranial pressure is above safe levels, the patient is put on a ventilator to increase the breathing rate. The more breaths a patient takes, the more carbon dioxide is blown off, helping to shrink cerebral blood vessels and thus reduce intracranial pressure. Drugs, such as mannitol, help draw water away from the brain.

Doctors also now can identify blood clots with CT and MRI scans and remove the clots before a patient completely deteriorates.

An estimated 250,000 individuals are living with spinal cord injury in the U.S. Some 11,000 new injuries are reported annually and are caused mostly by motor vehicle accidents, violence and falls. Economic costs approach $10 billion a year.

Researchers have found that people who suffer spinal cord injuries become less severely paralyzed if they receive high intravenous doses of a commonly used steroid drug within eight hours after injury. The drug *methylprednisolone* appears to help regardless of how severely the spinal cord is injured, and in some cases, makes the difference between a patient being confined to a wheelchair and being able to walk. Building on this knowledge, researchers hope to decipher the precise order of chemical reactions that leads to damage.

Scientists have known that following a complete injury to the spinal cord, animals can regain the ability to bear their weight and walk at various speeds on a treadmill belt. More recently, scientists have recognized that the level of this recovery is dependent to a large degree on whether these tasks are practiced—that is, trained—after injury. It appears that humans with spinal cord injury also respond to training interventions.

Anxiety disorders

The most widespread mental illnesses, anxiety disorders annually affect an estimated 12.6 percent of the adult population, or 24.8 million Americans. They include phobias, panic disorder and agoraphobia and obsessive-compulsive disorder (OCD). Some can keep people completely housebound or, as in the case of panic disorder, contribute to suicide.

In OCD, people become trapped, often for many years, in repetitive thoughts and behaviors, which they recognize as groundless but cannot stop, such as repeatedly washing hands, or checking doors or stoves. The illness is estimated to affect 3.8 million Americans annually. Social learning and genetics may play a role in developing the disorder. But positron emission tomography (PET) scans reveal abnormalities in both cortical and deep areas of the brain, suggesting a biological component as well.

Scientists recently discovered that certain breeds of large dogs that develop *acral lick syndrome*, severely sore paws through compulsive licking, respond to the serotonergic antidepressant clomipramine, which was the first effective treatment developed for OCD in people.

Serotonergic antidepressants, especially the tricyclics, clomipramine and serotonin reuptake inhibitors are effective in treating OCD. A specialized type of behavioral intervention, *exposure and response prevention*, is also effective in many patients.

Panic disorder, which affects 2.4 million Americans annually, usually starts "out of the blue." Patients experience an overwhelming sense of impending doom, accompanied by sweating, weakness, dizziness and shortness of breath. With repeated attacks, patients may develop anxiety in anticipation of another attack and avoid public settings where attacks might occur. Untreated, their lives may constrict until they develop *agoraphobia*, or the fear of crowds.

The recent discovery of brain receptors for the benzodiazepine anti-anxiety drugs has sparked research to identify the brain's own anti-anxiety chemical messengers. This finding may lead to ways to regulate this brain system and correct its possible defects in panic disorder. PET scans reveal that during such attacks, the tip of the brain's temporal lobe is unusually active compared with controls. When normal people expect to receive a shock to the finger, the same general area is activated.

The serotonin reuptake inhibitors, cognitive behavior therapy, or a combination are now the first choice treatments of panic disorder. Tricyclic antidepressants, MAO inhibitors and high-potency benzodiazepines also are effective.

Schizophrenia

Marked by disturbances in thinking, emotional reactions and social behavior, schizophrenia usually results in chronic illness and personality change. Delusions, hallucinations and thought disorder are common.

Affecting about one percent of the population or 2 million Americans a year, schizophrenia is disabling and costly. On a given day, these patients occupy up to 100,000 hospital beds. Annual costs total about $32.5 billion.

Schizophrenia is thought to reflect changes in the brain, possibly caused by disease or injury at the time of birth, or a genetic disposition that may be exacerbated by environmental stress. Brain systems using dopamine appear to be particularly involved. Brain scans and postmortem studies show abnormalities in some people with schizophrenia, such as enlarged ventricles (fluid-filled spaces) and reduced size of certain brain regions. PET scans taken during intellectual tasks show abnormal functioning in specific brain areas of persons with this illness.

The disorder usually begins in persons between the ages of 15 and 25. Some patients fully recover following treatment, but most continue to have moderate or severe symptoms, particularly in response to stress. About 15 percent of patients return to normal life after a single episode; 60 percent will have intermittent episodes throughout their lives; another 25 percent will not recover their ability to live as independent adults.

After a long search for an effective antipsychotic medication, scientists synthesized the drug *chlorpromazine* during the late 1940s. By the 1950s, it was found useful for treating psychotic states and later became a mainstay of drug treatment.

Since then a large number of agents similar to chlorpromazine have been developed. When given as long-acting injections, these drugs reduce some symptoms and aid patients' readiness for adjustment back into the community. However, chronic use may cause abnormal muscle movements and tremors in some patients. Safer treatments are being sought.

Thus far, most drugs are successful in treating hallucinations and thought disorder. *Clozapine*, acts somewhat differently from other antipsychotics. It treats the approximately 30 percent of patients who are not helped by conventional medications. However, the drug can induce a potentially fatal blood disorder, *agranulocytosis*, in about one percent of patients. To prevent this disorder, patients must take regular weekly to biweekly blood tests, a precaution that makes the use of the drug very costly. Several new antipsychotics—*risperidone, olanzapine* and *sertindole*—are now available. They do not involve risk of angranulocytosis but may have other side effects.

Neurological AIDS

By the end of 2000, some 448,000 deaths and up to 774,000 infections from acquired immune deficiency syndrome (AIDS) had occurred in the U.S. This is dwarfed by the more than 21.8 million deaths and 58 million infections identified worldwide.

While the principal target of *human immunodeficiency virus* (HIV) is the immune system, the nervous system also may be profoundly affected. Some 20 percent to 40 percent of patients with full-blown AIDS also develop clinically significant dementia that includes movement impairment. Those affected have mental problems ranging from mild difficulty with concentration or coordination to progressive, fatal dementia.

Despite advances in treating other aspects of the disease, AIDS dementia remains a mystery. Most current hypotheses center on an *indirect* effect of HIV infection related to secreted viral products or cell-coded signal molecules called *cytokines*. Nonetheless, HIV infection appears to be the prime mover in this disorder since antiviral treatment may prevent or reverse this condition in some patients.

Experts believe that serious neurologic symptoms are uncommon early in AIDS infection. But later, patients develop difficulty with concentration and memory and experience general slowing of their mental processes. At the same time, patients may develop leg weakness and a loss of balance. Imaging techniques, such as CT and MRI, show that the brains in these patients have undergone some shrinkage. The examination of brain cells under a microscope suggests that abnormalities are present principally in subcortical areas. Neurons in the cortex also may be altered.

Recent studies indicate that highly active combination antiretroviral treatment ('cocktails' of three or more drugs active against HIV) is effective in reducing the incidence of AIDS dementia. Such treatment also can effectively reverse the cognitive abnormalities attributed to brain HIV infection.

Despite this remarkable progress, some patients develop these problems and fail to respond to treatment, thus requiring additional approaches to prevention and treatment of these symptoms as well as the common peripheral neuropathy that can afflict those with AIDS.

Multiple sclerosis

The most common central nervous system disease of young adults after epilepsy, multiple sclerosis (MS) is a life-long ailment of unknown origin that affects more than 300,000 Americans. MS is diagnosed in individuals who are mainly between the ages of 20 and 50, with two of three cases occurring in women. MS results in earning losses of about $2 billion annually for families with MS.

Although a cause has yet to be found, MS is thought to be an autoimmune disease in which the body's natural defenses act against the *myelin* in the central nervous system as though it were foreign tissue. In MS, myelin is destroyed and replaced by scars of hardened "sclerotic" patches of tissue. Such lesions are called "plaques," and appear in multiple places within the central nervous system. This can be compared to a loss of insulating material around an electrical wire, which interferes with the transmission of signals. Some nerve fibers are actually cut in association with the loss of myelin.

Siblings of people with MS are 10 to 15 times more likely than others to be afflicted by the disorder. In addition, the disease is five times more prevalent in temperate zones, such as the Northern United States and Northern Europe, than it is in the tropics. Thus, genetic and environmental factors are probably involved in the cause. An infection acquired during the first 15 years of life may be responsible for triggering the disease in a genetically susceptible individual.

The most common symptoms are blurred vision, awkward gait, numbness and fatigue. These can occur singly or in combination, vary in intensity and last from several weeks to months. In some patients, symptoms include slurred speech, weakness, loss of coordination, uncontrollable tremors, loss of bladder control, memory problems, depression and paralysis. Muscle spasticity can affect balance and coordination, causing pain and involuntary jerking movement—and, if untreated, can create *contractures* or the "freezing" of a joint that prevents movement.

MS cannot be cured at present, but several medications control relapsing forms of MS. A wide range of medications and therapies are available to control symptoms such as spasticity, pain, fatigue, mood swings and bladder, bowel or sexual dysfunctions. Steroids, which have been used in MS for three decades, effectively shorten attacks and speed recovery from MS-related optic nerve inflammation. Promising new agents to control MS or to alleviate its symptoms are in clinical trials.

Down syndrome

Down syndrome, the most frequently occurring chromosomal abnormality, appears in one out of every 800 to 1,000 babies born. It occurs when an extra copy of chromosome 21 or part of its long arm is present in the egg or, less commonly, the sperm, at the time of conception. It is not known why this error in cell division occurs. It is not linked to any environmental or behavioral factors, either before or during pregnancy.

This disorder is associated with approximately 50 physical and developmental characteristics. An individual with Down syndrome is likely to possess, to varying degrees, some of these characteristics. They include mild to moderate mental retardation, low muscle tone, an upward slant to the eyes, a flat facial profile, an enlarged tongue and an increased risk of congenital heart defects, respiratory problems and obstructed digestive tracts.

The risk of having a child with this syndrome increases with the age of the mother. At age 35, the risk is about one in 365 births. At age 40, it is one in 110. However, it is important to note that the average age of women who give birth to children with Down syndrome is 28. This is because younger women are giving birth more often.

Scientists are moving closer to understanding the role that the genes on chromosome 21 play in Down syndrome.

Prenatal screening tests, such as the Triple Screen and Alpha-fetaprotein Plus, can accurately detect about 60 percent of fetuses with Down syndrome.

Babies with Down syndrome will develop much like typical children, but at a somewhat slower rate. They will learn to sit, walk, talk and toilet train, just like their peers. Early intervention programs can begin shortly after birth and can help foster an infant's development.

Down syndrome patients have been able to have longer and fuller lives, thanks to medical advances and a greater understanding of the potential of those with this condition. Individuals with Down syndrome are being educated in their neighborhood schools, participating in community activities and finding rewarding employment and relationships.

Although there is no cure or means of preventing Down syndrome, scientists are moving closer to understanding the role that the genes on chromosome 21 play in a person's development. Once this mystery is understood, they hope to decode the biochemical processes that occur in Down syndrome and treat or cure this disorder.

Huntington's disease

Affecting some 30,000 Americans and placing another 150,000 at risk, Huntington's disease (HD) is now considered one of the most common hereditary brain disorders. The disease that killed folk singer Woody Guthrie in 1967 progresses slowly over a ten to 20-year period and eventually robs the affected individual of the ability to walk, talk, think and reason. HD usually appears between the ages of 30 and 50. It affects both the basal ganglia that controls coordination and the brain cortex, which serves as the center for thought, perception and memory.

The most recognizable symptoms include involuntary jerking movements of the limbs, torso and facial muscles. These are often accompanied by mood swings, depression, irritability, slurred speech and clumsiness. As the disease progresses, common symptoms include difficulty swallowing, unsteady gait, loss of balance, impaired reasoning and memory problems. Eventually, the individual becomes totally dependent on others for care, with death often due to pneumonia, heart failure or another complication.

Diagnosis consists of a detailed clinical examination and family history. Brain scans may be helpful. The identification in 1993 of the gene that causes HD has simplified genetic testing, which can be used to help confirm a diagnosis. However, HD researchers and genetic counselors have established specific protocols for predictive testing to ensure that the psychological and social consequences of a positive or negative result are understood. Predictive testing is available only for adults, though children under 18 may be tested to confirm a diagnosis of juvenile onset HD. Prenatal testing may be performed. The ethical issues of testing must be considered and the individual adequately informed, because there is no effective treatment or cure.

The HD mutation is an expanded triplet repeat in the HD gene—a kind of molecular stutter in the DNA. This abnormal gene codes for an abnormal protein called *huntingtin*. The huntingtin protein, whose normal function is still obscure, is widely distributed in the brain and appears to be associated with the intracellular machinery involved in the transport of proteins. But the cause of HD probably involves a gain of a new and toxic function. Cell and transgenic animal models can replicate many features of the disease and are now being used to test new theories and therapies. Many researchers hope that transplanted or resident stem cells may one day be able to replace the neurons that have been lost to the disease.

Tourette syndrome

One of the most common and least understood neurobiological disorders, Tourette syndrome (TS) is a genetic condition that affects an estimated one in 500 Americans, roughly 200,000 people. Males are affected three to four times as frequently as females.

Symptoms usually appear between the ages of four and eight, but in rare cases may emerge as late as age 18. The symptoms include motor and vocal tics that are repetitive, involuntary movements or utterances that are rapid and sudden. The types of tics may change frequently, and increase or decrease in severity over time. Generally, this disorder lasts a lifetime, but one-third of patients may experience a remission or decrease in symptoms as they get older. Most people with TS do not require medication; their symptoms are mild and do not affect functioning.

The disorder seems to result from a hypersensitivity of dopamine receptors. Another neurotransmitter, serotonin, also has been implicated. The most effective drugs for control of movements, such as haloperidol, act by blocking the overactive system. Other symptoms, such as obsessive-compulsive traits and attention deficit disorder, often require treatment with other classes of drugs that act on serotonin.

The neuroleptic drugs *haloperidol* and *pimozide* have been the mainstays of treatment. They are not perfect medications, however, because they can cause disturbing side effects—abnormal involuntary movements, stiffness of the face and limbs, or sedation and weight gain in some patients. Recently, newer medications have been found effective in some patients.

Brain tumors

Although brain tumors are not always *malignant*—a condition that spreads and becomes potentially lethal—these growths are always serious because they can cause pressure in the brain and compression of nearby structures, interfering with normal brain activity.

Primary brain tumors arise within the brain while secondary brain tumors spread from other parts of the body through the bloodstream. For tumors starting in the brain, about 60 percent of which are malignant, the cause is unknown. Tumors that begin as cancer elsewhere and spread to the brain are always malignant.

The incidence of primary brain tumors is about 12 per 100,000 population. About 36,000 new cases occur in the United States annually. Because of difficulties diagnosing and classifying brain tumors, exact statistics on secondary tumors are unknown.

Symptoms vary according to location and size. The compression of brain tissue or nerve tracts, as well as expansion of the tumor, can cause symptoms such as seizures, headaches, muscle weakness, loss of vision or other sensory problems and speech difficulties. An expanding tumor can increase pressure within the skull, causing headache, vomiting, visual disturbances and impaired mental functioning. Brain tumors are diagnosed with MRI and CT scanning.

Surgery is a common treatment if the tumor is accessible and vital structures will not be disturbed. Radiation is used to stop a tumor's growth or cause it to shrink. Chemotherapy destroys tumor cells that may remain after surgery and radiation. Steroid drugs relieve swelling and other symptoms. *Immunotherapy* uses the body's own immune system against the tumor. Promising areas of research include *bioengineered genes*, monoclonal antibodies that attach to specifically targeted cells; *growth factors; angiogenesis inhibitors* and *differentiation therapies; targeted toxins;* and *tumor vaccines*.

Amyotrophic lateral sclerosis

This fatal disorder strikes 5,000 Americans annually with 50 percent of patients dying within three to five years of diagnosis. It is the most common disorder within a group of diseases affecting movement and costs Americans some $300 million annually.

Commonly known as Lou Gehrig's disease, amyotrophic lateral sclerosis (ALS) destroys neurons that control voluntary muscle movements, such as walking. For reasons that are not understood, brain and spinal motor neurons in the spinal cord begin to disintegrate. Because signals from the brain are not carried by these damaged nerves to the body, the muscles begin to weaken and deteriorate from the lack of stimulation and use.

The first signs of progressive paralysis are usually seen in the hands and feet. They include leg weakness, walking difficulty, and clumsiness of the hands when washing and dressing. Eventually, almost all muscles under voluntary control, including those of the respiratory system, are affected. Despite the paralysis, the mind and the senses remain intact. Death is usually caused by respiratory failure or pneumonia.

No specific test identifies ALS; but muscle biopsies, blood studies, electrical tests of muscle activity, CT and MRI scans and X-rays of the spinal cord help identify the disease and rule out other disorders. Still, diagnosis is often difficult because its causes remain unknown. Potential causes include glutamate toxicity, oxidative stress, factors in the environment and an autoimmune response in which the body's defenses turn against body tissue.

In about 90 percent of cases, ALS is sporadic, arising in individuals with no known family history of the disorder. In the other 10 percent of cases, it is *familial*—transmitted to family members because of a gene defect.

Scientists recently found a gene responsible for one form of ALS. Mutations in the gene that codes for *super oxide dismutase* located on chromosome 21 were linked to the presence of this disorder. Scientists believe that whatever they learn from studying the gene will have relevance for understanding other forms of motor neuron disease.

Once diagnosed, physical therapy and rehabilitation methods help strengthen unused muscles. Various drugs can ease specific problems like twitching and muscle weakness, but there is no cure. An antiglutamate drug modestly slows down the disease. Additional drugs are now under study. Protecting or regenerating motor neurons using nerve growth factors and stem cells may someday provide significant hope for patients.

New diagnostic methods

Many of the recent advances in understanding the brain are due to the development of techniques that allow scientists to directly monitor neurons throughout the body.

Electrophysiological recordings trace brain electrical activity in response to a specific external stimulus. In this method, electrodes placed in specific parts of the brain—depending on which sensory system is being tested—make recordings that are then processed by a computer. The computer makes an analysis based on the time lapse between stimulus and response. It then extracts this information from background activity.

Following the discovery that material is transported within neurons, methods have been developed to visualize activity and precisely track fiber connections within the nervous system. This can be done by injecting a radioactive amino acid into the brain of an experimental animal; the animal is killed a few hours later; and then the presence of radioactive cells is visualized on film. In another technique, the enzyme horseradish peroxidase is injected and taken up by nerve fibers that can be later identified under a microscope.

These and other methods have resulted in many advances in knowledge about the workings of the nervous system and are still useful today. New methods, safely applicable to humans, promise to give even more precise information, particularly about the point of origin of disorders such as epilepsy.

Imaging techniques

Positron emission tomography (PET) This method of measuring brain function is based on the detection of radioactivity emitted when positrons, positively charged particles, undergo radioactive decay in the brain. Substances labeled with positron-emitting radionuclides are used to produce three-dimensional PET images that reflect blood flow as well as metabolic and chemical activity in the brain.

So far, PET studies have helped scientists understand more about how drugs affect the brain and what happens during learning, language and certain brain disorders—such as stroke and Parkinson's disease. Within the next few years, PET could enable scientists to identify the biochemical nature of neurological and mental disorders and determine how well therapy is working in patients. For instance, depression produces very marked changes in the brain as seen by PET. Knowing the location of these changes helps researchers understand better the causes of depression and monitor the effectiveness of specific treatments.

Another technique, *single photon emission computed tomography* (SPECT), is similar to PET but its pictures are not as detailed. SPECT is much less expensive than PET because the tracers it uses have a longer half-life and do not require a nearby accelerator to produce them.

Magnetic resonance imaging (MRI) Providing a high-quality, three-dimensional image of organs and structures inside the body without X-rays or other radiation, MRI images are unsurpassed in anatomical detail and may reveal minute changes that occur with time.

MRI is expected to tell scientists when structural abnormalities first appear in the course of a disease, how they affect subsequent development and precisely how their progression correlates with mental and emotional aspects of a disorder.

During the 15-minute MRI imaging procedure, a patient lies inside a massive, hollow, cylindrical magnet and is exposed to a powerful, steady magnetic field. The protons of the body's hydrogen atoms, especially those in water and fat, normally point randomly in different directions, but in a very strong magnetic field (many times the earth's magnetic field) they line up parallel to each other like rows of tiny bar magnets. If the hydrogen nuclei are then knocked out of alignment by a strong pulse of radio waves, they produce a detectable radio signal as they fall back into alignment.

Magnetic coils in the machine detect these signals and a computer changes them into an image based on different types of body tissue. Tissue that contains a lot of water and fat produces a bright image; tissue that contains little or no water, such as bone, appears black. (The image is similar to that produced by CT scanning, but MRI generally gives much greater contrast between normal and abnormal tissues.)

MRI allows images to be constructed in any plane and is particularly valuable in studying the brain and spinal cord. It reveals tumors rapidly and vividly, indicating their precise extent. MRI provides early evidence of potential damage from stroke, thus allowing physicians to administer proper treatments early.

Magnetic resonance spectroscopy (MRS), a technique related to MRI which uses the same machinery but examines molecular composition and metabolic processes, rather than anatomy, also holds great promise to provide insights into how the brain works. By measuring the molecular and metabolic changes that occur in the brain, MRS has already provided new information on brain development and aging, Alzheimer's disease, schizophrenia, autism and stroke. Because this method is noninvasive, it is ideally suited to study the natural course of a disease or its response to treatment.

Functional magnetic resonance imaging (fMRI) Another exciting recent development in imaging is fMRI. This technique measures brain activity under resting and activated conditions. It combines the high spatial resolution, noninvasive imaging of brain anatomy offered by standard MRI with a strategy for detecting changes in blood oxygenation levels driven by neuronal activity. This technique allows for more detailed maps of brain areas underlying human mental activities in health and disease. To date, fMRI has been applied to the study of various functions of the brain ranging from primary sensory responses to cognitive activities. While the exact origin of the signal changes found in fMRI is still under debate, the success of fMRI in numerous studies has clearly demonstrated its great potential.

Magnetoencephalography (MEG) One of the latest advances in scanners, MEG reveals the source of weak magnetic fields emitted by neurons. An array of cylinder-shaped sensors monitor the magnetic field pattern near the patient's head to determine the positions and strengths of activity in various regions of the brain. In contrast with other imaging techniques, MEG can characterize rapidly changing patterns of neural activity with millisecond resolution and provide a quantitative measure of its strength for individual subjects. Moreover, by presenting stimuli at various rates, it is possible to determine how long the neural activation is sustained in the diverse brain areas that respond.

One of the most exciting developments in imaging is the combined use of information from fMRI and MEG. The former provides detailed information about the areas of brain activity in a particular task whereas MEG tells researcher and physician when they become active. The combined use of this information allows a much more precise understanding of how the brain works in health and disease.

Gene diagnosis

The inherited blueprint for all human characteristics, genes consist of short sections of *deoxyribonucleic acid* (DNA), the long, spiraling, helix structure found on the 23 pairs of *chromosomes* in the nucleus of every human cell.

New gene diagnosis techniques now make it possible to find the chromosomal location of genes responsible for neurologic and psychiatric diseases and to identify structural changes in these genes that are responsible for causing disease.

This information is useful for identifying individuals who carry faulty genes and thereby improving diagnosis; for understanding the precise cause of diseases in order to improve methods of prevention and treatment; and for evaluating the malignancy and susceptibility of certain tumors.

So far, scientists have identified defective genes for more than 50 neurological disorders and the chromosomal location of the defect in up to 100. Prenatal or carrier tests exist for many of the most prevalent of these illnesses.

Scientists have tracked down the gene on chromosome 4 that goes awry in Huntington's patients. The defect is an expansion of a CAG repeat. CAG is the genetic code for the amino acid glutamine, and the expanded repeat results in a long string of glutamines within the protein. This expansion appears to alter the protein's function. Scientists have found that the size of the expanded repeat in an individual is predictive of Huntington's disease. Other neurodegenerative disorders have been identified as due to expanded CAG repeats in other genes. The mechanisms by which these expansions caused adult onset neurodegeneration is the focus of intense research.

Sometimes patients with single gene disorders are found to have a chromosomal abnormality—a deletion or break in the DNA sequence of the gene—that can lead scientists to a more accurate position of the disease gene. This is the case with some abnormalities found on the X-chromosome in patients with *Duchenne muscular dystrophy* and on chromosome 13 in patients with *inherited retinoblastoma*, a rare childhood eye tumor that can lead to blindness and other cancers.

Gene mapping has led to the localization on chromosome 21 of the gene coding the beta amyloid precursor protein that is abnormally cut to form the smaller peptide, beta amyloid. It is this peptide that accumulates in the senile plaques that clog the brains of patients with Alzheimer's disease. This discovery shed light on the reason why individuals with Down syndrome (Trisomy 21) invariably accumulate amyloid deposits: they make too much amyloid as a consequence of having three copies instead of two copies of this gene. Mutations in this gene have recently been shown to underlie Alzheimer's in a distinct subset of these patients.

Several other genetic factors have been identified in Alzheimer's disease, including genes for two proteins, presenilin 1 and presenilin 2, located on chromosomes 14 and 1. A risk factor for late onset Alzheimer's is the gene for the apolipoprotein E protein located on chromosome 19.

Gene mapping has enabled doctors to diagnose *Fragile X mental retardation*, the most common cause of inherited mental retardation. Scientists now believe they have identified this gene.

Several groups of scientists are investigating whether there are genetic components to schizophrenia, manic depression and alcoholism, but their findings are not yet conclusive.

Overall, the characterizations of the structure and function of individual genes causing diseases of the brain and nervous system are in the early stages. Factors that determine variations in the genetic expression of a single-gene abnormality—such as what contributes to the early or late start or severity of a disorder—are still unknown.

Scientists also are studying the genes in *mitochondria*, structures found *outside* the cell nucleus that have their own genome and are responsible for the production of energy used by the cell. Recently, different mutations in mitochondrial genes were found to cause several rare neurological disorders. Some scientists speculate that an inheritable variation in mitochondrial DNA may play a role in diseases such as Alzheimer's, Parkinson's and some childhood diseases of the nervous system.

THE CELL

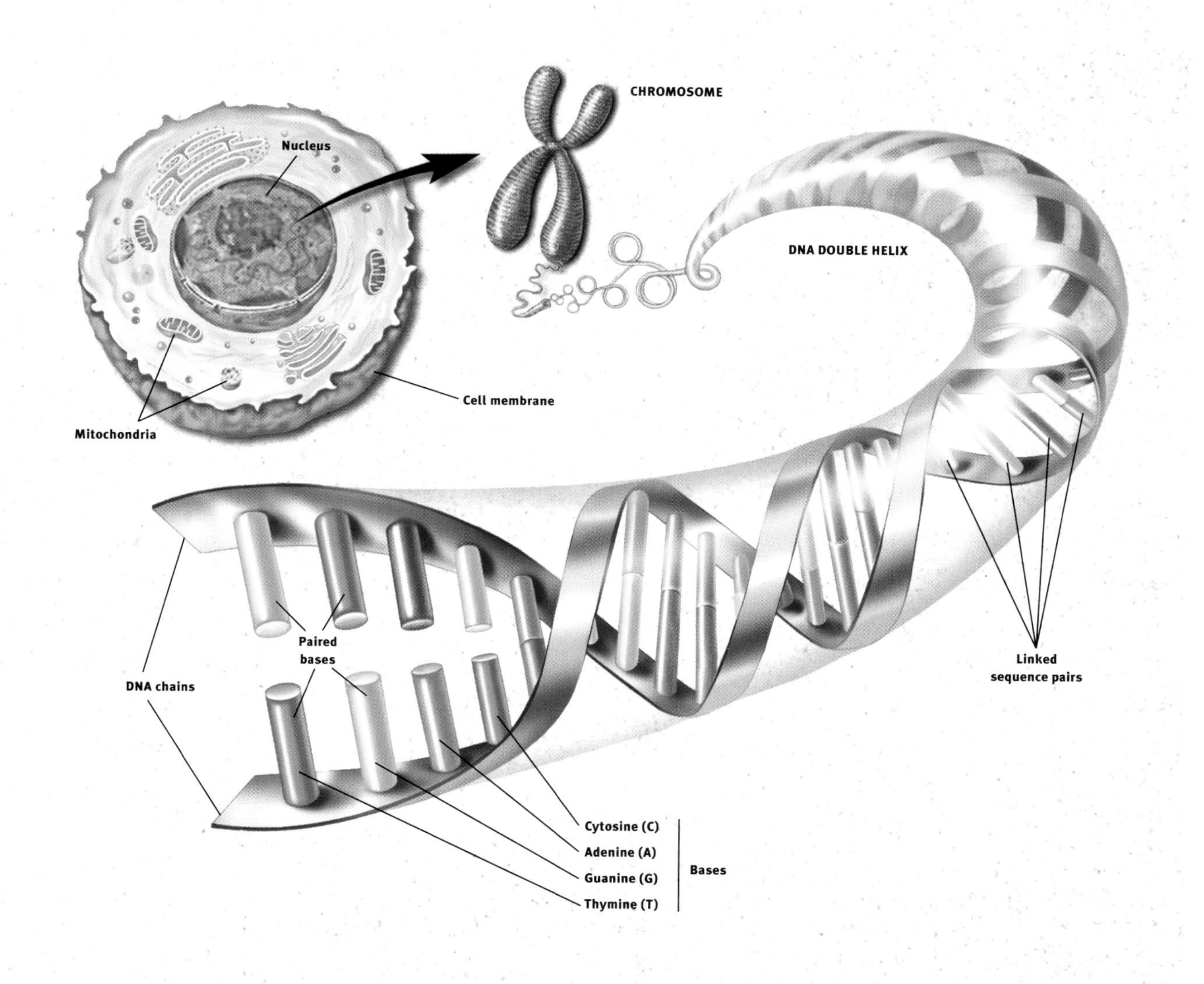

ANATOMY OF A GENE. Within the nucleus of every human cell, two long, threadlike strands of DNA encode the instructions for making all the proteins necessary for life. Each cell holds more than 50,000 different genes found on 23 paired chromosomes of tightly coiled DNA. Each strand of DNA bears four types of coding molecules—adenine (A), cytosine (C), guanine (G), and thymine (T). The sequence of coding molecules in a gene (segment of DNA) is the code for protein manufacture.

Potential therapies

New drugs. Most drugs used today were developed using trial-and-error techniques that often do not reveal why a drug produces a particular effect. But the expanding knowledge gained from the new methods of molecular biology—the ability to make a receptor gene and determine its structure—make it possible to design safer and more effective drugs.

In a test tube, the potency of an agent can be determined by how well it binds to a receptor. A scientist then can vary the drug's structure to enhance its action on the receptor. Thus, subsequent generations of drugs can be designed to interact with the receptors more efficiently, producing higher potency and fewer side effects.

While this "rational drug design" holds promise for developing drugs for conditions ranging from stroke and migraine headaches to depression and anxiety, it will take considerable effort to clarify the role of the different receptors in these disorders.

Trophic factors

One result of basic neuroscience research involves the discovery of numerous survival or *trophic factors* found in the brain that control the development and survival of specific groups of neurons. Once the specific actions of these molecules and their receptors are identified and their genes cloned, procedures can be developed to modify trophic factor-regulated function in ways that might be useful in the treatment of neurological disorders.

Already, researchers have demonstrated the possible value of at least one of these factors, *nerve growth factor* (NGF). Infused into the brains of rats, NGF prevented cell death and stimulated the regeneration and sprouting of damaged neurons that are known to die in Alzheimer's disease. When aged animals with learning and memory impairments were treated with NGF, scientists found that these animals were able to remember a maze task as well as healthy aged rats.

Recently, several new factors have been identified and are beginning to be studied. They are potentially useful for therapy, but scientists must first understand how they may influence neurons. Alzheimer's, Parkinson's and Lou Gehrig's diseases may be treated in the future with trophic factors or their genes.

Because the destruction of neurons that use acetylcholine is one feature of Alzheimer's disease, any substance that can prevent this destruction is an important topic of research. NGF, which can do this, also holds promise for slowing the memory deficits associated with normal aging.

Once a trophic factor for a particular cell is found, copies of the factor can be genetically targeted to the area of the brain where this type of cell has died. The treatment may not cure a disease but could improve symptoms and delay progression.

In an interesting twist on growth factor therapy, researchers for the first time demonstrated that a neutralization of inhibitory molecules can help repair damaged nerve fiber tracts in the spinal cord. Using antibodies to Nogo-A, a protein that inhibits nerve regeneration, Swiss researchers succeeded in getting nerves of damaged spinal cords in rats to regrow. Treated rats showed large improvements in their ability to walk after spinal cord damage.

In these experiments, scientists cut one of the major groups of nerve fiber tracts in the spinal cord that connect the spinal cord and the brain. When an antibody, directed against the factor Nogo-A, was administered to the spinal cords or the brains of adult rats, "massive sprouting" of nerve fibers occurred where the spinal cord had been cut. Within two to three weeks, neurons grew to the lower level of the spinal cord and in some animals along its whole length. In untreated spinal cord-injured rats, the maximum distance of nerve regrowth rarely exceeded one-tenth of an inch. This research could eventually have clinical implications for spinal cord or brain-damaged people.

Cell and gene therapy

Researchers throughout the world are pursuing a variety of new ways to repair or replace healthy neurons and other cells in the brain. Most of the experimental approaches are still being worked out in animals and cannot be considered real therapies at this time.

Scientists have identified an *embryonic neuronal stem cell*—an unspecialized cell that gives rise to cells with specific func-

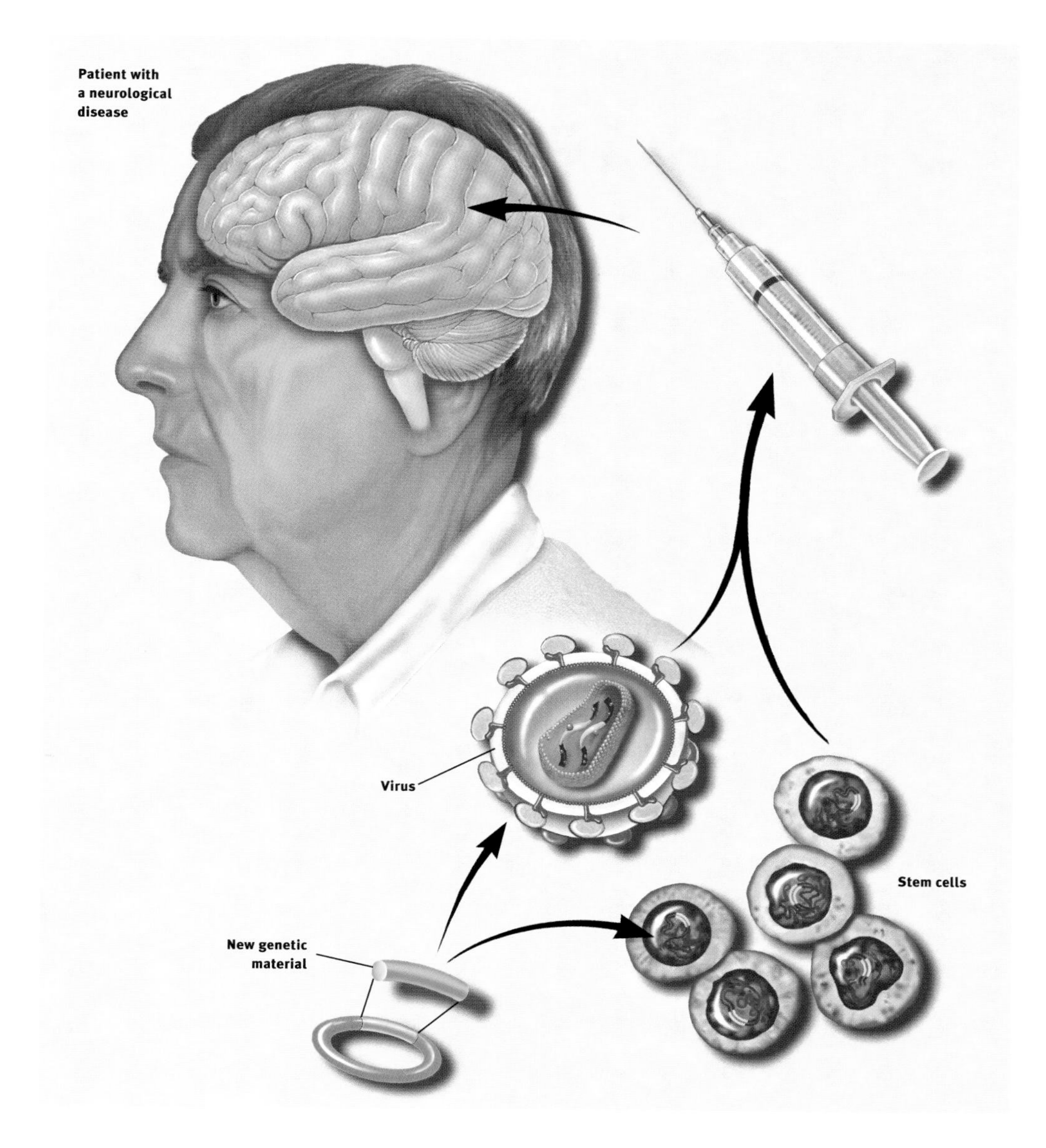

CELL AND GENE THERAPY. In potential therapy techniques, scientists plan to insert genetic material for a beneficial neurotransmitter or trophic factor into stem cells or a virus. The cells or virus are then put into a syringe and injected into the patient where they will produce the beneficial molecule and, it is hoped, improve symptoms.

tions. They have located this type of cell in the brain and spinal cord of embryonic and adult mice that can be stimulated to divide by known proteins, epidermal growth factor and fibroblast growth factor. The stem cells can continuously produce all three major cell types of the brain—neurons; *astrocytes*, the cells that nourish and protect neurons; and *oligodendrocytes*, the cells that surround axons and allow them to conduct their signals efficiently. Someday their production abilities may become useful for replacing missing neurons. A very similar stem cell also has been discovered in the adult nervous system in various kinds of tissue, raising the possibility that these stem cells can be pharmacologically directed to replace damaged neurons.

In other work, researchers are studying a variety of viruses that may ultimately be used to act as "Trojan horses" to carry therapeutic genes to the brain to correct nervous system diseases. The viruses include *herpes simplex type 1 virus (HSV), adenovirus, lentivirus, adeno-associated virus* and others naturally attracted to neurons. All have been found to be capable of being modified to carry new genes to cells in tissue culture and in the rodent central nervous system. HSV and adenovirus vectors have also been evaluated in early-stage human trials for treating brain tumors.

In one gene therapy experiment, scientists created an animal model of Parkinson's disease (PD) in rhesus monkeys. One week later, these monkeys received injections of the glial cell-derived neurotrophic factor (GDNF) gene into the striatum and substantia nigra using a lentiviral vector system. The nigrostriatal system is the main brain area affected by PD. The injections reversed the motor deficits seen on a clinical rating scale and a hand-reach task for up to three months. PET scans showed that these animals displayed marked increases in measures of dopamine, a chemical that is deficient in patients. Postmortem studies revealed a comprehensive protection in striatal dopamine as well as the number of nigrostriatal neurons. The results support the concept that lentiviral delivery of GDNF may provide neuroprotection for patients with early PD.

Glossary

ACETYLCHOLINE A neurotransmitter in both the brain, where it regulates memory, and in the peripheral nervous system, where it controls the actions of skeletal and smooth muscle.

ACTION POTENTIAL This occurs when a neuron is activated and temporarily reverses the electrical state of its interior membrane from negative to positive. This electrical charge travels along the axon to the neuron's terminal where it triggers the release of a neurotransmitter.

ADRENAL CORTEX An endocrine organ that secretes corticosteroids for metabolic functions; for example, in response to stress.

ADRENAL MEDULLA An endocrine organ that secretes epinephrine and norepinephrine in concert with the activation of the sympathetic nervous system; for example, in response to stress.

AGONIST A neurotransmitter, a drug or other molecule that stimulates receptors to produce a desired reaction.

ALZHEIMER'S DISEASE The major cause of dementia most prevalent in the elderly, it inflicts enormous human financial cost on society. The disease is characterized by death of neurons in the hippocampus, cerebral cortex and other brain regions.

AMINO ACID TRANSMITTERS The most prevalent neurotransmitters in the brain, these include glutamate and aspartate, which have excitatory actions, and glycine and gamma-amino butyric acid (GABA), which have inhibitory actions.

AMYGDALA A structure in the forebrain that is an important component of the limbic system and plays a central role in emotional learning.

ANDROGENS Sex steroid hormones, including testosterone, found in higher levels in males than females. They are responsible for male sexual maturation.

ANTAGONIST A drug or other molecule that blocks receptors. Antagonists inhibit the effects of agonists.

APHASIA Disturbance in language comprehension or production, often as a result of a stroke.

AUDITORY NERVE A bundle of nerve fibers extending from the cochlea of the ear to the brain, which contains two branches: the cochlear nerve that transmits sound information and the vestibular nerve that relays information related to balance.

AUTONOMIC NERVOUS SYSTEM A part of the peripheral nervous system responsible for regulating the activity of internal organs. It includes the sympathetic and parasympathetic nervous systems.

AXON The fiberlike extension of a neuron by which the cell sends information to target cells.

BASAL GANGLIA Clusters of neurons, which include the caudate nucleus, putamen, globus pallidus and substantia nigra, located deep in the brain that play an important role in movement. Cell death in the substantia nigra contributes to Parkinson's disease.

BRAINSTEM The major route by which the forebrain sends information to and receives information from the spinal cord and peripheral nerves. The brainstem controls, among other things, respiration and regulation of heart rhythms.

BROCA'S AREA The brain region located in the frontal lobe of the left hemisphere that is important for the production of speech.

CATECHOLAMINES The neurotransmitters dopamine, epinephrine and norepinephrine that are active both in the brain and the peripheral sympathetic nervous system. These three molecules have certain structural similarities and are part of a larger class of neurotransmitters known as monoamines.

CEREBELLUM A large structure located at the roof of the hindbrain that helps control movement by making connections to the pons, medulla, spinal cord and thalamus. It also may be involved in aspects of motor learning.

CEREBRAL CORTEX The outermost layer of the cerebral hemispheres of the brain. It is responsible for all forms of conscious experience, including perception, emotion, thought and planning.

CEREBRAL HEMISPHERES The two specialized halves of the brain. The left hemisphere is specialized for speech, writing, language and calculation; the right hemisphere is specialized for spatial abilities, face recognition in vision and some aspects of music perception and production.

CEREBROSPINAL FLUID A liquid found within the ventricles of the brain and the central canal of the spinal cord.

CHOLECYSTOKININ A hormone released from the lining of the stomach during the early stages of digestion which acts as a powerful suppressant of normal eating. It also is found in the brain.

CIRCADIAN RHYTHM A cycle of behavior or physiological change lasting approximately 24 hours.

CLASSICAL CONDITIONING Learning in which a stimulus that naturally produces a specific response (unconditioned stimulus) is repeatedly paired with a neutral stimulus (conditioned stimulus). As a result, the conditioned stimulus can evoke a response similar to that of the unconditioned stimulus.

COCHLEA A snail-shaped, fluid-filled organ of the inner ear responsible for transducing motion into neurotransmission to produce an auditory sensation.

COGNITION The process or processes by which an organism gains knowledge or becomes aware of events or objects in its environment and uses that knowledge for comprehension and problem-solving.

CONE A primary receptor cell for vision located in the retina. The cone is sensitive to color and used primarily for daytime vision.

CORPUS CALLOSUM The large bundle of nerve fibers linking the left and right cerebral hemispheres.

CORTISOL A hormone manufactured by the adrenal cortex. In humans, cortisol is secreted in greatest quantities before dawn, readying the body for the activities of the coming day.

DEPRESSION A mental disorder characterized by depressed mood and abnormalities in sleep, appetite and energy level.

DENDRITE A tree-like extension of the neuron cell body. Along with the cell body, it receives information from other neurons.

DOPAMINE A catecholamine neurotransmitter known to have multiple functions depending on where it acts. Dopamine-containing neurons in the substantia nigra of the brainstem project to the caudate nucleus and are destroyed in Parkinson's victims. Dopamine is thought to regulate emotional responses and play a role in schizophrenia and drug abuse.

DORSAL HORN An area of the spinal cord where many nerve fibers from peripheral pain receptors meet other ascending and descending nerve fibers.

DRUG ADDICTION Loss of control over drug intake or compulsive seeking and taking of drugs, despite adverse consequences.

ENDOCRINE ORGAN An organ that secretes a hormone directly into the bloodstream to regulate cellular activity of certain other organs.

ENDORPHINS Neurotransmitters produced in the brain that generate cellular and behavioral effects like those of morphine.

EPILEPSY A disorder characterized by repeated seizures, which are caused by abnormal excitation of large groups of neurons in various brain regions. Epilepsy can be treated with many types of anticonvulsant medications.

EPINEPHRINE A hormone, released by the adrenal medulla and specialized sites in the brain, that acts with norepinephrine to affect the sympathetic division of the autonomic nervous system. Sometimes called adrenaline.

ESTROGENS A group of sex hormones found more abundantly in females than males. They are responsible for female sexual maturation and other functions.

EVOKED POTENTIALS A measure of the brain's electrical activity in response to sensory stimuli. This is obtained by placing electrodes on the surface of the scalp (or more rarely, inside the head), repeatedly administering a stimulus and then using a computer to average the results.

EXCITATION A change in the electrical state of a neuron that is associated with an enhanced probability of action potentials.

FOLLICLE-STIMULATING HORMONE A hormone released by the pituitary gland that stimulates the production of sperm in the male and growth of the follicle (which produces the egg) in the female.

FOREBRAIN The largest division of the brain, which includes the cerebral cortex and basal ganglia. The forebrain is credited with the highest intellectual functions.

FRONTAL LOBE One of the four divisions (parietal, temporal, occipital) of each hemisphere of the cerebral cortex. The frontal lobe has a role in controlling movement and in the planning and coordinating of behavior.

GAMMA-AMINO BUTYRIC ACID (GABA) An amino acid transmitter in the brain whose primary function is to inhibit the firing of neurons.

GLIA Specialized cells that nourish and support neurons.

GLUTAMATE An amino acid neurotransmitter that acts to excite neurons. Glutamate stimulates N-methyl-D-aspartate (NMDA) receptors that have been implicated in activities ranging from learning and memory to development and specification of nerve contacts in a developing animal. Stimulation of NMDA receptors may promote beneficial changes, while overstimulation may be a cause of nerve cell damage or death in neurological trauma and stroke.

GONAD Primary sex gland: testis in the male and ovary in the female.

GROWTH CONE A distinctive structure at the growing end of most axons. It is the site where new material is added to the axon.

HIPPOCAMPUS A seahorse-shaped structure located within the brain and considered an important part of the limbic system. It functions in learning, memory and emotion.

HORMONES Chemical messengers secreted by endocrine glands to regulate the activity of target cells. They play a role in sexual development, calcium and bone metabolism, growth and many other activities.

HUNTINGTON'S DISEASE A movement disorder caused by death of neurons in the basal ganglia and other brain regions. It is characterized by abnormal movements called chorea—sudden, jerky movements without purpose.

HYPOTHALAMUS A complex brain structure composed of many nuclei with various functions. These include regulating the activities of internal organs, monitoring information from the autonomic nervous system, controlling the pituitary gland and regulating sleep and appetite.

INHIBITION In reference to neurons, it is a synaptic message that prevents the recipient cell from firing.

IONS Electrically charged atoms or molecules.

LIMBIC SYSTEM A group of brain structures—including the amygdala, hippocampus, septum, basal ganglia and others—that help regulate the expression of emotion and emotional memory.

LONG-TERM MEMORY The final phase of memory in which information storage may last from hours to a lifetime.

MANIA A mental disorder characterized by excessive excitement, exalted feelings, elevated mood, psychomotor overactivity and overproduction of ideas. It may be associated with psychosis; for example, delusions of grandeur.

MELATONIN Produced from serotonin, melatonin is released by the pineal gland into the bloodstream. Melatonin affects physiological changes related to time and lighting cycles.

MEMORY CONSOLIDATION The physical and psychological changes that take place as the brain organizes and restructures information in order to make it a permanent part of memory.

METABOLISM The sum of all physical and chemical changes that take place within an organism and all energy transformations that occur within living cells.

MIDBRAIN The most anterior segment of the brainstem. Along with the pons and medulla, the midbrain is involved in many functions, including regulation of heart rate, respiration, pain perception and movement.

MITOCHONDRIA Small cylindrical particles inside cells that provide energy for the cell by converting sugar and oxygen into special energy molecules, called ATP.

MONOAMINE OXIDASE (MAO) The brain and liver enzyme that normally breaks down the catecholamines norepinephrine, dopamine, and epinephrine and other monosomines such as serotonin.

MOTOR NEURON A neuron that carries information from the central nervous system to muscle.

MYASTHENIA GRAVIS A disease in which acetylcholine receptors on muscle cells are destroyed so that muscles can no longer respond to the acetylcholine signal in order to contract. Symptoms include muscular weakness and progressively more common bouts of fatigue. The disease's cause is unknown but is more common in females than in males and usually strikes between the ages of 20 and 50.

MYELIN Compact fatty material that surrounds and insulates axons of some neurons.

NERVE GROWTH FACTOR A substance whose role is to guide neuronal growth during embryonic development, especially in the peripheral nervous system. Nerve growth factor also probably helps sustain neurons in the adult.

NEURON Nerve cell. It is specialized for the transmission of information and characterized by long fibrous projections called axons and shorter, branch-like projections called dendrites.

NEUROTRANSMITTER A chemical released by neurons at a synapse for the purpose of relaying information to other neurons via receptors.

NOCICEPTORS In animals, nerve endings that signal the sensation of pain. In humans, they are called pain receptors.

NOREPINEPHRINE A catecholamine neurotransmitter, produced both in the brain and in the peripheral nervous system. It is involved in arousal, and regulation of sleep, mood and blood pressure.

OCCIPITAL LOBE One of the four subdivisions of the cerebral cortex. The occipital lobe plays a role in processing visual information.

ORGANELLES Small structures within a cell that maintain the cells and do the cells' work.

PARASYMPATHETIC NERVOUS SYSTEM A branch of the autonomic nervous system concerned with the conservation of the body's energy and resources during relaxed states.

PARIETAL LOBE One of the four subdivisions of the cerebral cortex. The parietal lobe plays a role in sensory processes, attention and language.

PARKINSON'S DISEASE A movement disorder caused by death of dopamine neurons in the substantia nigra located in the midbrain. Symptoms include tremor, shuffling gait and general paucity of movement.

PEPTIDES Chains of amino acids that can function as neurotransmitters or hormones.

PERIPHERAL NERVOUS SYSTEM A division of the nervous system consisting of all nerves that are not part of the brain or spinal cord.

PHOSPHORYLATION A process that modifies the properties of neurons by acting on an ion channel, neurotransmitter receptor or other regulatory protein. During phosphorylation, a phosphate molecule is placed on a protein and results in the activation or inactivation of the protein. Phosphorylation is believed to be a necessary step in allowing some neurotransmitters to act and is often the result of second messenger activity.

PINEAL GLAND An endocrine organ found in the brain. In some animals, the pineal gland serves as a light-influenced biological clock.

PITUITARY GLAND An endocrine organ closely linked with the hypothalamus. In humans, the gland is composed of two lobes and secretes a number of hormones that regulate the activity of other endocrine organs in the body.

PONS A part of the hindbrain that, with other brain structures, controls respiration and regulates heart rhythms. The pons is a major route by which the forebrain sends information to and receives information from the spinal cord and peripheral nervous system.

PSYCHOSIS A severe symptom of mental disorders characterized by an inability to perceive reality. It can occur in many conditions, including schizophrenia, mania, depression and drug-induced states.

RECEPTOR CELL A specialized sensory cell designed to pick up and transmit sensory information.

RECEPTOR MOLECULE A specific protein on the surface or inside of a cell with a characteristic chemical and physical structure. Many neurotransmitters and hormones exert their effects by binding to receptors on cells.

REUPTAKE A process by which released neurotransmitters are absorbed for subsequent reuse.

ROD A sensory neuron located in the periphery of the retina. The rod is sensitive to light of low intensity and specialized for nighttime vision.

SCHIZOPHRENIA A chronic mental disorder characterized by psychosis (e.g., hallucinations and delusions), flattened emotions and impaired cognitive function.

SECOND MESSENGERS Substances that trigger communications among different parts of a neuron. These chemicals play a role in the manufacture and release of neurotransmitters, intracellular movements, carbohydrate metabolism and processes of growth and development. The messengers direct effects on the genetic material of cells may lead to long-term alterations of behavior, such as memory and drug addiction.

SEROTONIN A monoamine neurotransmitter believed to play many roles, including, but not limited to, temperature regulation, sensory perception and the onset of sleep. Neurons using serotonin as a transmitter are found in the brain and in the gut. A number of antidepressant drugs are targeted to brain serotonin systems.

SHORT-TERM MEMORY A phase of memory in which a limited amount of information may be held for several seconds to minutes.

STIMULUS An environmental event capable of being detected by sensory receptors.

STROKE The third largest cause of death in America, stroke is an impeded blood supply to the brain. Stroke can be caused by a rupture of a blood vessel wall, an obstruction of blood flow caused by a clot or other material or by pressure on a blood vessel (as by a tumor). Deprived of oxygen, which is carried by blood, nerve cells in the affected area cannot function and die. Thus, the part of the body controlled by those cells cannot function either. Stroke can result in loss of consciousness and death.

SYMPATHETIC NERVOUS SYSTEM A branch of the autonomic nervous system responsible for mobilizing the body's energy and resources during times of stress and arousal.

SYNAPSE A gap between two neurons that functions as the site of information transfer from one neuron to another.

TEMPORAL LOBE One of the four major subdivisions of each hemisphere of the cerebral cortex. The temporal lobe functions in auditory perception, speech and complex visual perceptions.

THALAMUS A structure consisting of two egg-shaped masses of nerve tissue, each about the size of a walnut, deep within the brain. The key relay station for sensory information flowing into the brain, the thalamus filters out only information of particular importance from the mass of signals entering the brain.

VENTRICLES Of the four ventricles, comparatively large spaces filled with cerebrospinal fluid, three are located in the forebrain and one in the brainstem. The lateral ventricles, the two largest, are symmetrically placed above the brainstem, one in each hemisphere.

WERNICKE'S AREA A brain region responsible for the comprehension of language and the production of meaningful speech.

Index

Numbers in **bold** refer to illustrations.

11 Dupont Circle, NW, Suite 500
Washington, DC 20036 USA

Telephone (202) 462-6688
www.sfn.org

To acquire additional copies of this book, please visit our Web site www.sfn.org, go to *Publications* and click on *Brain Facts.*

The Society for Neuroscience

Editor: Joseph Carey, Senior Director, Communications & Public Affairs
Science writer: Leah Ariniello
Researcher: Mary McComb

Produced by Meadows Design Office Incorporated, Washington, DC
www.mdomedia.com

Creative Director and Designer: Marc Alain Meadows
Production Assistants: Ching Huang Ooi, Nancy Bratton

Illustrator: Lydia V. Kibiuk, Baltimore, Maryland

Printed and bound in China by Everbest Printing Company

Fourth edition

06 05 04 03 02 5 4 3 2 1

ISBN 0-916110-00-1